CHROMOSOME ATLAS:

FISH, AMPHIBIANS, REPTILES AND BIRDS

VOLUME 2

CHROMOSOME ATLAS:
FISH, AMPHIBIANS, REPTILES AND BIRDS

VOLUME 2

Coordinating Editors

KURT BENIRSCHKE
T. C. HSU

Editors

MARIA LUIZA BEÇAK
WILLY BEÇAK
FRANKLIN L. ROBERTS
ROBERT N. SHOFFNER
E. PETER VOLPE

Springer Science+Business Media, LLC

Originally published by Springer-Verlag New York Inc. in 1973.
Softcover reprint of the hardcover 1st edition 1973

Library of Congress Catalog Card Number 73-166079

ISBN 978-3-642-49090-3 ISBN 978-3-642-65751-1 (eBook)
DOI 10.1007/978-3-642-65751-1

Introduction

Since the publication of the first Atlas of Mammalian Chromosomes in 1967 the continuous compilation of mammalian karyotypes has become a useful instrument in cytologic and taxonomic studies. Technical advances in preparing mitotic cells from nonmammalian vertebrates have since allowed a better comparison of taxa in fishes, amphibia, reptiles, and birds. In these fields the literature is also widely scattered; and it has become difficult to survey such information, published as well as unpublished, by nonspecialists. These were among the reasons for the new endeavor of compiling a chromosome atlas for nonmammalian vertebrates.

An annual publication is planned with presentation of between 10 and 15 karyotypes from each class. In this second volume, 52 species are presented. For convenience in future colation, the numbering system employs class abbreviations, viz., P-Pisces, Am-Amphibia, R-Reptilia, and Av-Aves. Within each class, the numbers are necessarily consecutive.

In general the karyotypes are laid out following the format employed in An Atlas of Mammalian Chromosomes. Whenever possible both sexes are represented, even though sexual chromosomal dimorphism is not (currently) evident. When the sex chromosomes are known, they are so indicated by conventional nomenclature (XX/XY or ZW/ZZ). In the karyotypes of birds the so-called microchromosomes are grouped together at the end without an attempt at complete enumeration, which is presently impossible. They are usually considered as acrocentrics, but a few are distinctly biarmed. The modal number is quoted as the most likely in these species, recognizing the difficulty in obtaining accurate counts of the microchromosomes.

Whenever possible, the number of specimens studied and their origin are given. Very old references are usually avoided. New references will again be furnished with each new Atlas. In Pisces, the nomenclature of Common and Scientific Names of Fishes, American Fisheries Society Publication, 1970 (3rd ed.) is employed.

The karyotypes displayed here come primarily from the files of the editors and reflect what has been available and of interest to them. In the future volumes a wider representation of species will be presented. It is hoped that investigators with karyotypes of animals belonging to these classes will contact individual editors and contribute their valuable material for the inclusion in the future volumes. Proper acknowledgment will be made in individual cases.

Coordinating Editors:

Kurt Benirschke
Department of Obstetrics & Gynecology
School of Medicine
University of California, San Diego
La Jolla, California 92037

T. C. Hsu
Section of Cell Biology
M. D. Anderson Hospital
Houston, Texas 77025

September, 1973

Editors:

Franklin L. Roberts
(Pisces)
Department of Zoology
University of Maine
Orono, Maine 04473

E. Peter Volpe
(Amphibia)
Department of Biology
Tulane University
New Orleans, Louisiana 70118

Maria Luiza Beçak
Willy Beçak
(Reptilia)
Serviço De Genética
Instituto Butantan
Caixa Postal 65
São Paulo, Brazil

Robert N. Shoffner
(Aves)
Department of Animal Science
University of Minnesota
St. Paul, Minnesota 55101

Contents, Volume 2

Corrigendum to Volume 1:

Cumulative Contents,
Volumes 1 & 2

Amphisbaena vermicularis Wagler (Worm lizard; Cobra de duas cabeças) R-25
Leposternon microcephalum Wagler (Cobra de duas cabeças) R-26

ANGUIDAE
Ophiodes striatus (Spix) (Glass snake; Cobra de vidro) R-28

BOIDAE
Boa constrictor amarali (Stull) (Boa constrictor, Jibóia) R-3
Epicrates cenchria crassus (Cope) (Rainbow boa, Salamanta) R-4
Eunectes murinus (Linnaeus) (Anaconda, Sucuri) R-5

COLUBRIDAE
Chironius bicarinatus (Wied) (Cipo snake, Cobra cipó) R-6
Clelia occipitolutea (Duméril) (Black snake Muçurana; Cobra preta) R-17
Drymarchon corais corais (Boie) (South American indigo snake; Papa-ovo; Papa-pinto) R-18
Liophis miliaris (Linnaeus) (Cobra d'agua) R-19
Mastigodryas bifossatus bifossatus (Raddi) (Panther snake; Jararacuçu do brejo; Nyakaniná) R-20
Philodryas olfersii olfersii (Lichtenstein) (Olfer's green snake, Cobra verde) R-7
Philodryas serra (Schlegel) (Cobra cipó) R-21
Spilotes pullatus anomalepis Bocourt (Chicken snake; Caninana; Caninana do papo amarelo; Yacaniña) R-22
Thamnodynastes strigatus (Günther) (Thamnodynastes) R-8
Tomodon dorsatus Duméril et Bibron (Corre campo) R-9
Xenodon merremii (Wagler) Merrem's xenodon, Boipeva) R-10
Xenodon neuwiedii Günther (Neuwied's snake; Quiriripitá) R-23

IGUANIDAE
Anolis carolinensis Cuvier (Chamaleon lizard) R-2
Tropidurus torquatus (Wied) (Calango) R-29

SCINCIDAE
Mabuya mabouya mabouya (Lacépède) (Mabuya; Lagartixa) R-27

VIPERIDAE
Bothrops alternatus Duméril (Urutu cruzeiro; Jararaca rabo de porco) R-16
Bothrops jararaca (Wied) (Jararaca) R-11
Bothrops jararacussu Lacerda (Jararacuçu) R-12
Bothrops moojeni Hoge (Moojen's fer de lance, Caiçaca) R-13
Crotalus durissus terrificus (Laurenti) (Neotropical rattlesnake, Cascavel) R-14

AVES
ANSERIFORMES
ANATIDAE
ANATINI
Anas acuta (Pintail duck) Av-25
Anas clypeata (Common shoveler) Av-26
Anas discors discors (Prairie blue-winged teal) Av-27
Anas platyrhynchos platyrhynchos (Mallard and its domesticated varieties) Av-28
Anas streptera (Gadwall) Av-29
AYTHYINI
Aythya affinis (Lesser scaup) Av-31
Aythya americana (Redhead) Av-32
Aythya valisineria (Canvasback) Av-30
CAIRININI
Aix galericulata (Mandarin duck) Av-21
Aix sponsa (Wood duck) Av-22
Cairina moschata (Muscovy duck) Av-23
DENDROCYGNINI
Dendrocygna bicolor (Fulvous tree duck) Av-24
MERGINI
Bucephala clangula americana (American goldeneye) Av-34
TADORNINI
Tadorna tadorna (Common shelduck) Av-33
APODIFORMES
TROCHILIDAE
Calypte anna (Anna's humming bird) Av-19
CHARADRIIFORMES
CHARADRIIDAE
Charadrius vociferus (Killdeer) Av-16
COLUMBIFORMES
COLUMBIDAE
Columba livia domestica (Common pigeon) Av-12
Streptopelia risoria (Ringneck dove) Av-13
Zenaidura macroura (Mourning dove) Av-14
GALLIFORMES
CRACIDAE
Mitu mitu (Curassow) Av-1
MELEAGRIDIDAE
Meleagris gallopavo (Turkey) Av-10
NUMIDIDAE
Numida meleagris (Guinea fowl) Av-11
PHASIANIDAE
Chrysolophus pictus (Golden pheasant) Av-2

Colinus virginianus (Bobwhite quail) Av-3
Collipepla squamata (Scaled quail) Av-4
Coturnix coturnix japonica (Japanese quail) Av-5
Gallus domesticus (Domestic fowl) Av-6
Lophortyx californicus (California quail) Av-7
Lophortyx gambelli (Gambell's quail) Av-20
Lophura swinhoei (Swinhoe's pheasant) Av-8
Phasianus colchicus (Chinese ringneck pheasant) Av-9

PASSERIFORMES

CORVIDAE

Corvus corax (Raven) Av-17

RHEIFORMES

RHEIDAE

Rhea americana (Rhea) Av-15

STRIGIFORMES

STRIGIDAE

Bubo virginianus (Great horned owl) Av-18

Order: PERCIFORMES

Family: PERCICHTHYIDAE

Morone americana (Gmelin)

(White perch)

2n = 48

Order: PERCIFORMES Family: PERCICHTHYIDAE

Morone americana (Gmelin) (White perch)

2n = 48

TYPES: Acrocentrics and subtelocentrics

The karyotype of 48 chromosomes in this species consists of one-arm chromosomes. Most have very short second arms, while a few show no evidence of second arms at all. In size, the chromosomes show a gradual gradation from largest to smallest, and hence cannot be easily divided into size groupings. Chromosomal sexual dimorphism has not been detected.

Both karyotypes are from primary cell cultures. The upper from a male set up with testicular and heart tissue, and the lower from a female set up with ovarian tissue. The specimens were collected in South Branch Lake, Penobscot County, Maine.

Morone americana (Gmelin) (White perch)

2n = 48

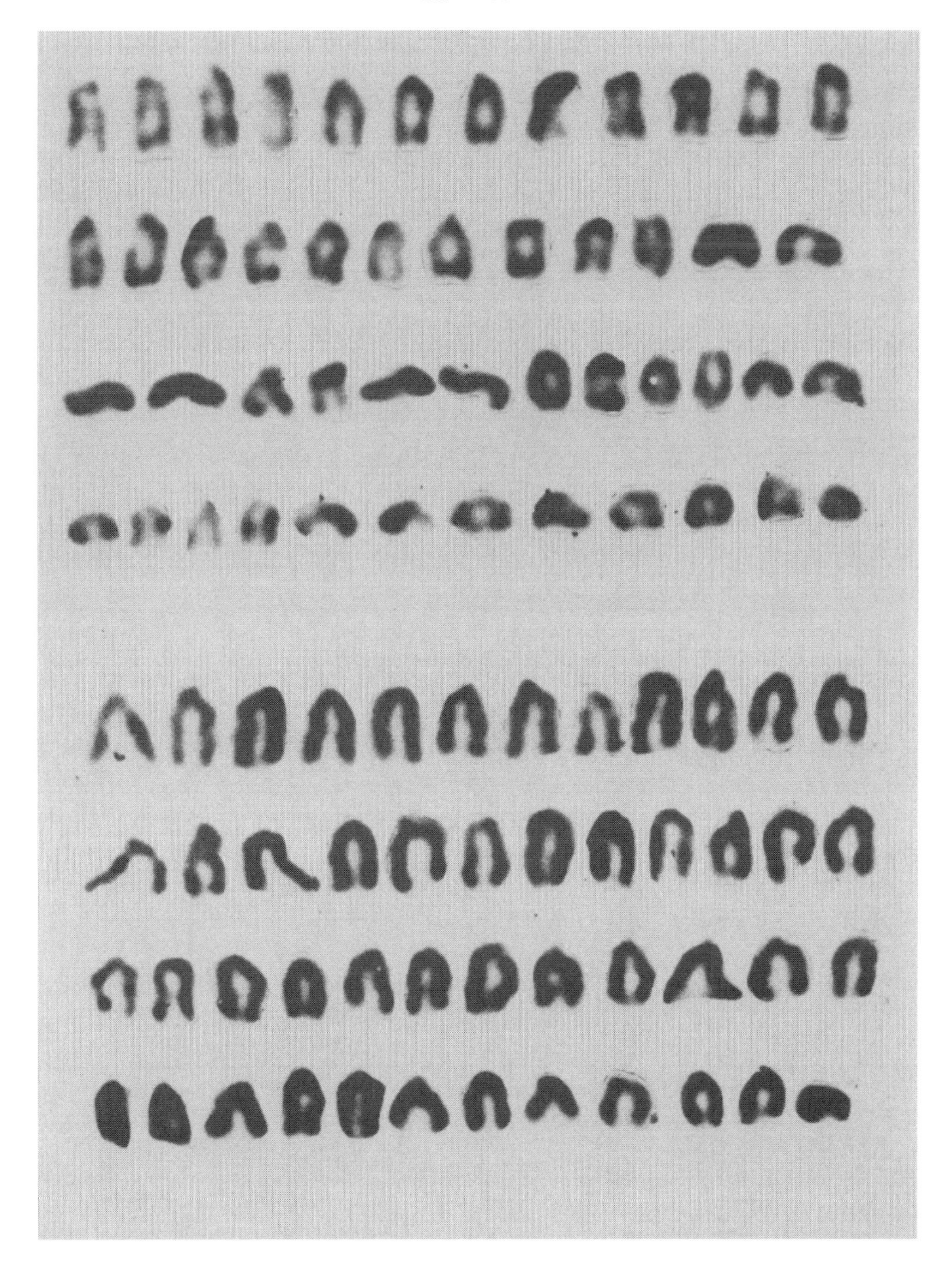

Order: PERCIFORMES

Family: PERCIDAE

Perca flavescens (Mitchill)

(Yellow perch)

2n = 48

Order: PERCIFORMES Family: PERCIDAE

Perca flavescens (Mitchill) (Yellow perch)

2n = 48

TYPES: Subtelocentrics

The karyotype of 48 chromosomes consists of one-arm chromosomes, all of which show distinct short arms. The chromosomes show gradation in size, with the exception of one pair which is distinctly smaller than the others. There is no clear evidence of sexual dimorphism.

The karyotypes were obtained from primary cell cultures. The upper is from a culture seeded with heart and testicular tissue from a male, while the lower is from an ovarian culture. The specimens were collected in South Branch Lake, Penobscot County, Maine.

Perca flavescens (Mitchill) (Yellow perch)

2n = 48

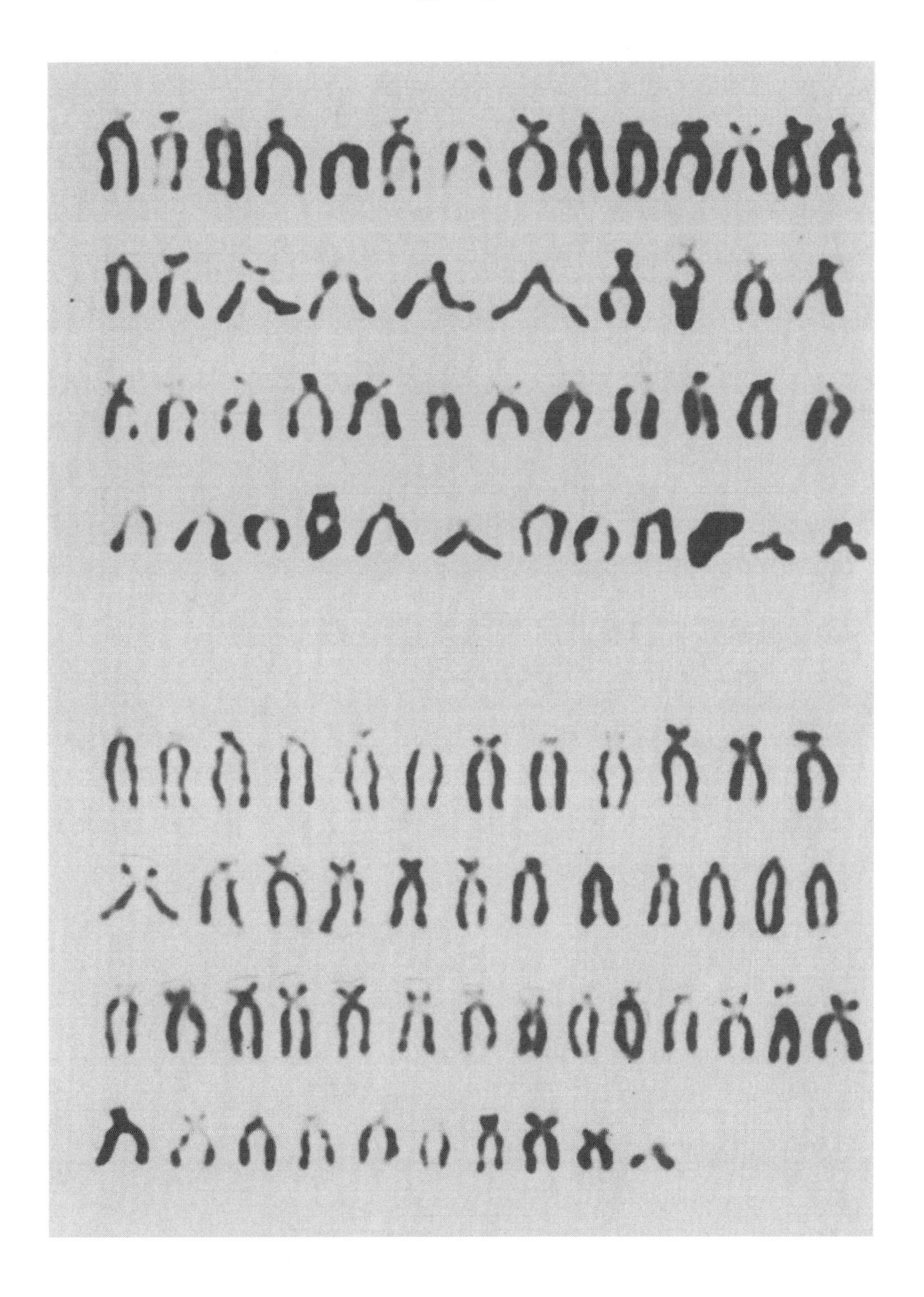

Order: ATHERINIFORMES

Family: CYPRINODONTIDAE

Megupsilon aporus (Miller and Walters)

2n = 47 (♂), 48 (♀)

Order: ATHERINIFORMES Family: CYPRINODONTIDAE

Megupsilon aporus Miller and Walters

2n = 47 (♂), 48 (♀)

TYPES: Metacentrics, acrocentrics and subtelocentrics

This species is quite unusual in showing chromosomal sexual dimorphism. The male possesses 47 chromosomes including a very large metacentric, while the female has a diploid complement of 48. The karyotype includes a medium-sized pair of metacentric chromosomes, and a graded series of one-arm chromosomes. The male metacentric sex chromosome is about four times as large as any of the other chromosomes in the karyotype.

The karyotypes were prepared from cells of the gill epithelium. The upper is a camera lucida drawing from a male, while the lower is a camera lucida drawing from a female. These were provided by Drs. T. Uyeno and Robert Miller, Museum of Zoology, University of Michigan, Ann Arbor, Michigan. The specimens were collected in Nuevo Leon, Mexico.

REFERENCES:

1) Uyeno, T. and Miller, R.: Multiple sex chromosomes in a mexican Cyprinodontid fish. Nature 231: 452, 1971.

2) Miller, R., and Walters, V.: A new genus of cyprinodontid fish from Neuvo Leon, Mexico. Contributions Sci., Natural History Museum, Los Angeles County. No. 233: 1, 1972.

Megupsilon aporus Miller and Walters

2n = 47 (♂), 48 (♀)

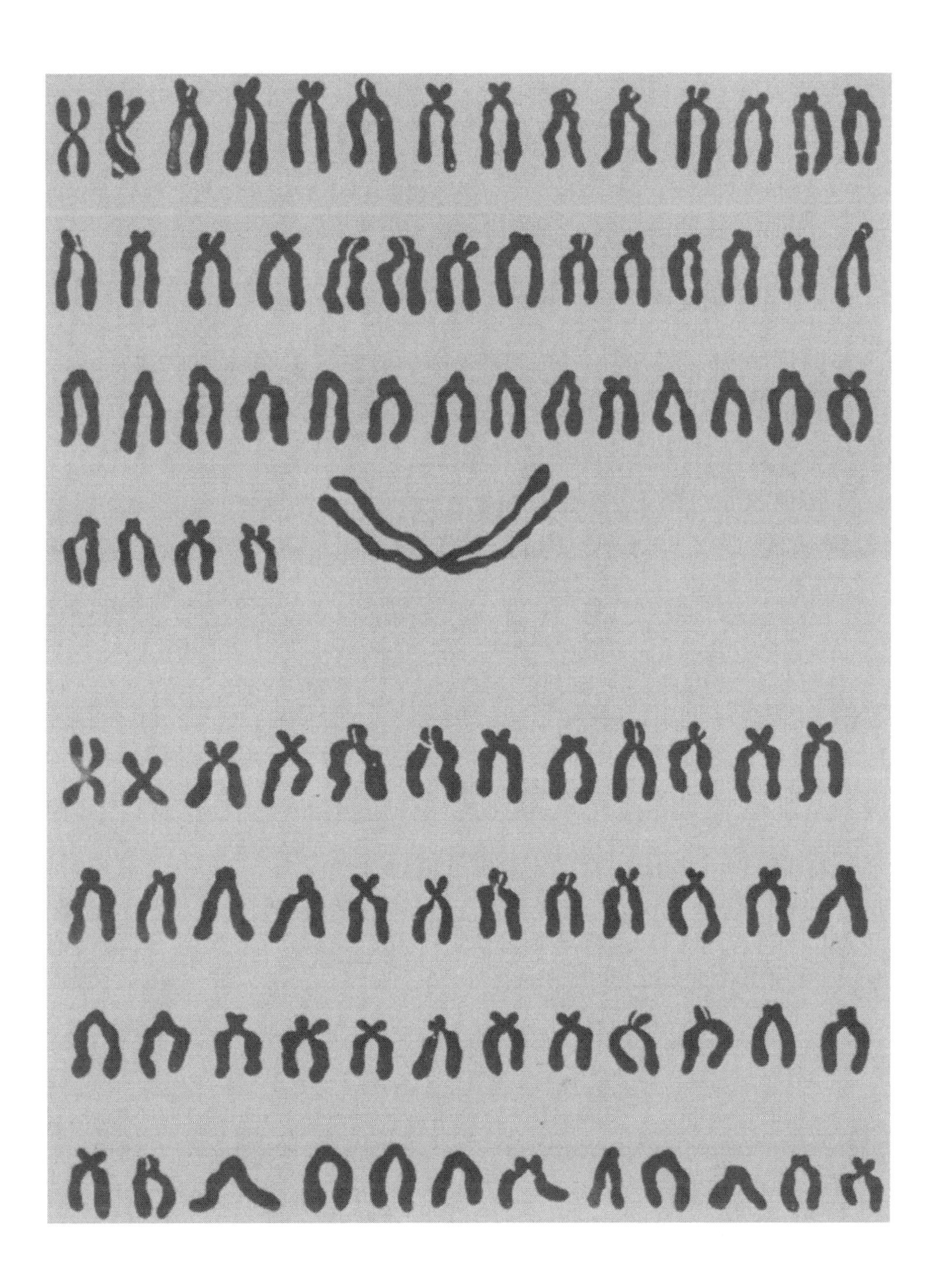

Order: SILURIFORMES

Family: ICTALURIDAE

Noturus gyrinus (Mitchill)

(Tadpole madton)

2n = 42

Order: SILURIFORMES Family: ICTALURIDAE

Noturus gyrinus (Mitchill) (Tadpole madton)

2n = 42

TYPES: Metacentrics and acrocentrics

This karyotype consists of 11 pairs of metacentric chromosomes showing a wide range in size and 10 pairs of subtelocentrics. Except for one relatively large pair, the subtelocentrics are quite small. There is no evidence of sexual dimorphism.

The upper karyotype is from a male while the lower is from a female. Both preparations were taken from gill filament epithelial cells.

Both karyotypes were provided by Catharine B. Levin, Department of Limnology, Academy of Natural Sciences, Philadelphia, Pennsylvania.

Noturus gyrinus (Mitchill) (Tadpole madton)

2n = 42

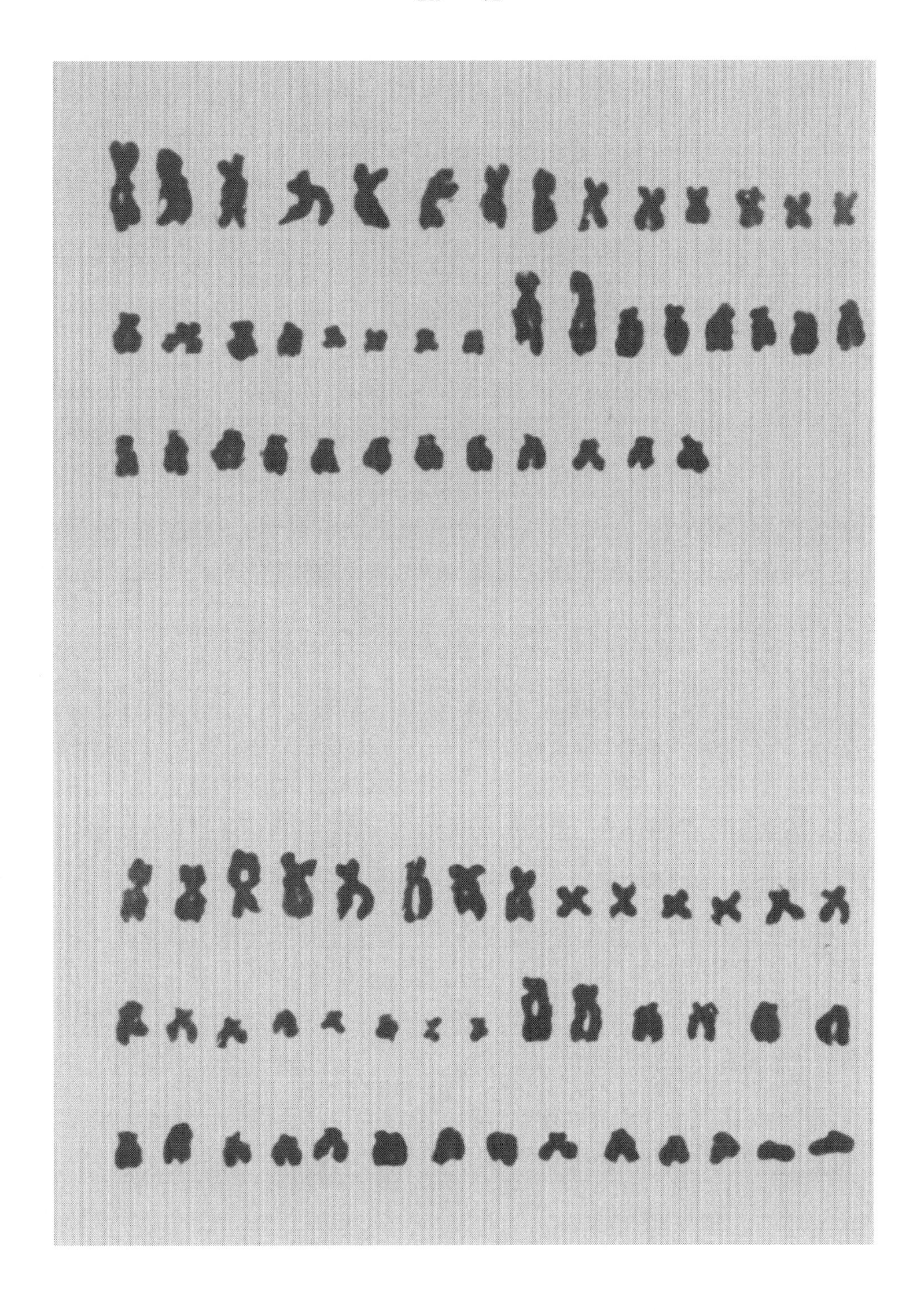

Order: SILURIFORMES

Family: ICTALURIDAE

Ictalurus nebulosus (Lesueur)

(Brown bullhead)

$2n = 60$

Order: SILURIFORMES Family: ICTALURIDAE

Ictalurus nebulosus (Lesueur) (Brown bullhead)

2n = 60

TYPES: Metacentrics, submetacentrics, acrocentrics and subtelocentrics

This karyotype consists of 60 chromosomes, most of which are subtelocentrics. There is a gradation both in chromosome size and in length of the shorter arm. The few metacentrics are mostly among the smaller chromosomes of the karyotype, a situation that is somewhat atypical for fishes. The arrangement of chromosomes on the plate is somewhat arbitrary, since clear divisions are non-existent. Chromosomal sexual dimorphism is not apparent.

These karyotypes were obtained from primary cell cultures. The upper was set up with cells from a male, while the lower consists of female cells. The specimens were collected in South Branch Lake, Penobscot County, Maine.

Ictalurus nebulosus (Lesueur) (Brown bullhead)

2n = 60

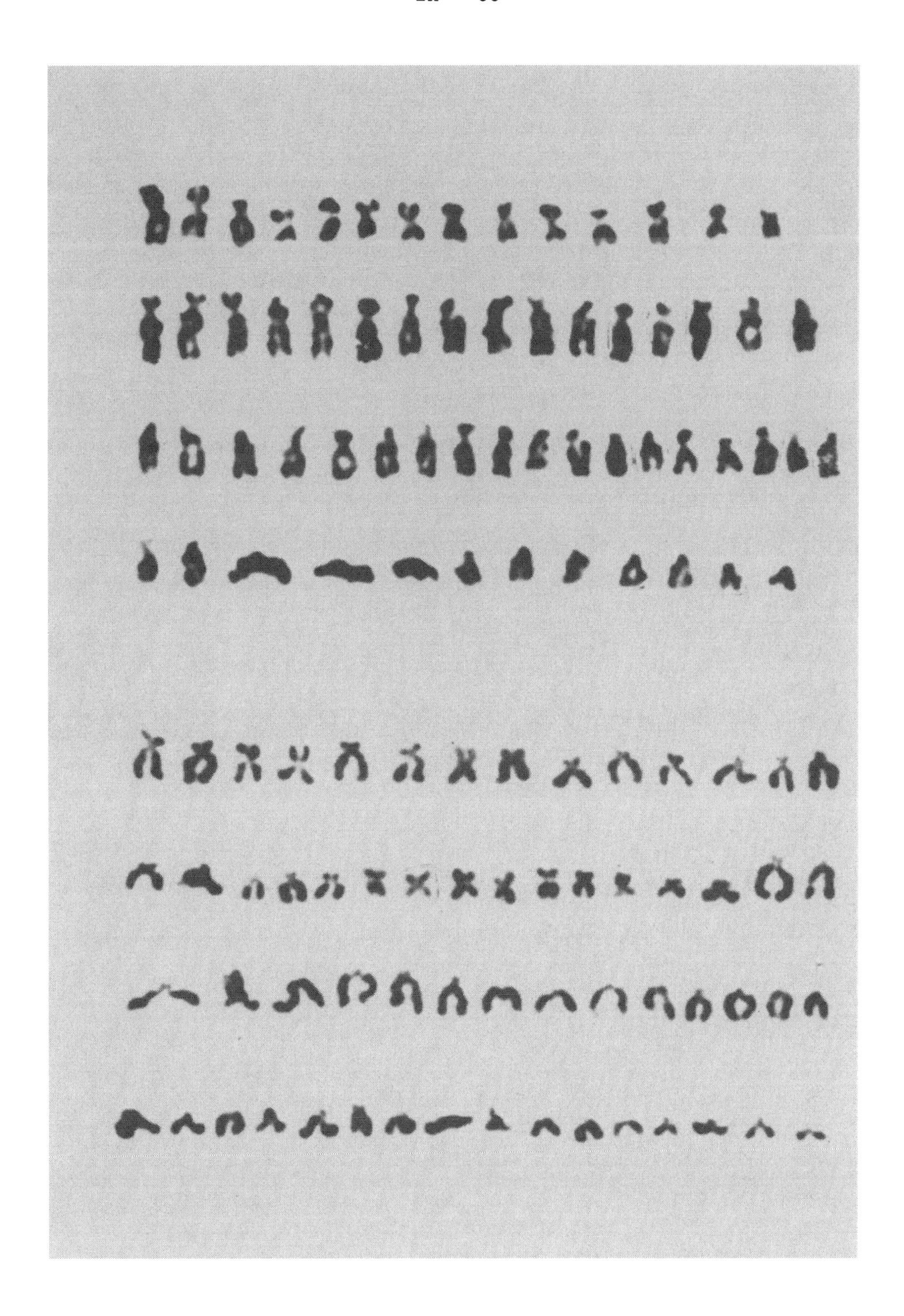

Order: CYPRINIFORMES

Family: CATOSTOMIDAE

Catostomus commersoni (Lacépède)

(White sucker)

2n = 100

Order: CYPRINIFORMES Family: CATOSTOMIDAE

Catostomus commersoni (Lacépède) (White sucker)

2n = 100

TYPES: Metacentrics, submetacentrics, acrocentrics, subtelocentrics

Of the total karyotype of 100 chromosomes, about 24 are metacentrics and submetacentrics, while the remainder are acrocentrics and subtelocentrics. In view of the likelihood that this species is tetraploid, it should be possible to arrange these chromosomes in groupings of four. An examination of the karyotype indicates that this might be possible but, unfortunately, the homologues are not sufficiently different from each other to be sure of such grouping. There is no evidence of chromosomal sexual dimorphism, and in fact such would not be expected in view of the probable tetraploid origin.

The upper karyotype is from a primary cell culture of testicular tissue and heart tissue from a male, while the lower is from a primary culture of ovarian tissue. The specimens were collected at South Branch Lake, Penobscot County, Maine.

REFERENCES:

1) Uyeno, T. and Smith, G.: Tetraploid origin of the karyotype of catastomid fishes. Science 175: 644, 1972.

Catostomus commersoni (Lacépède) (White sucker)

2n = 100

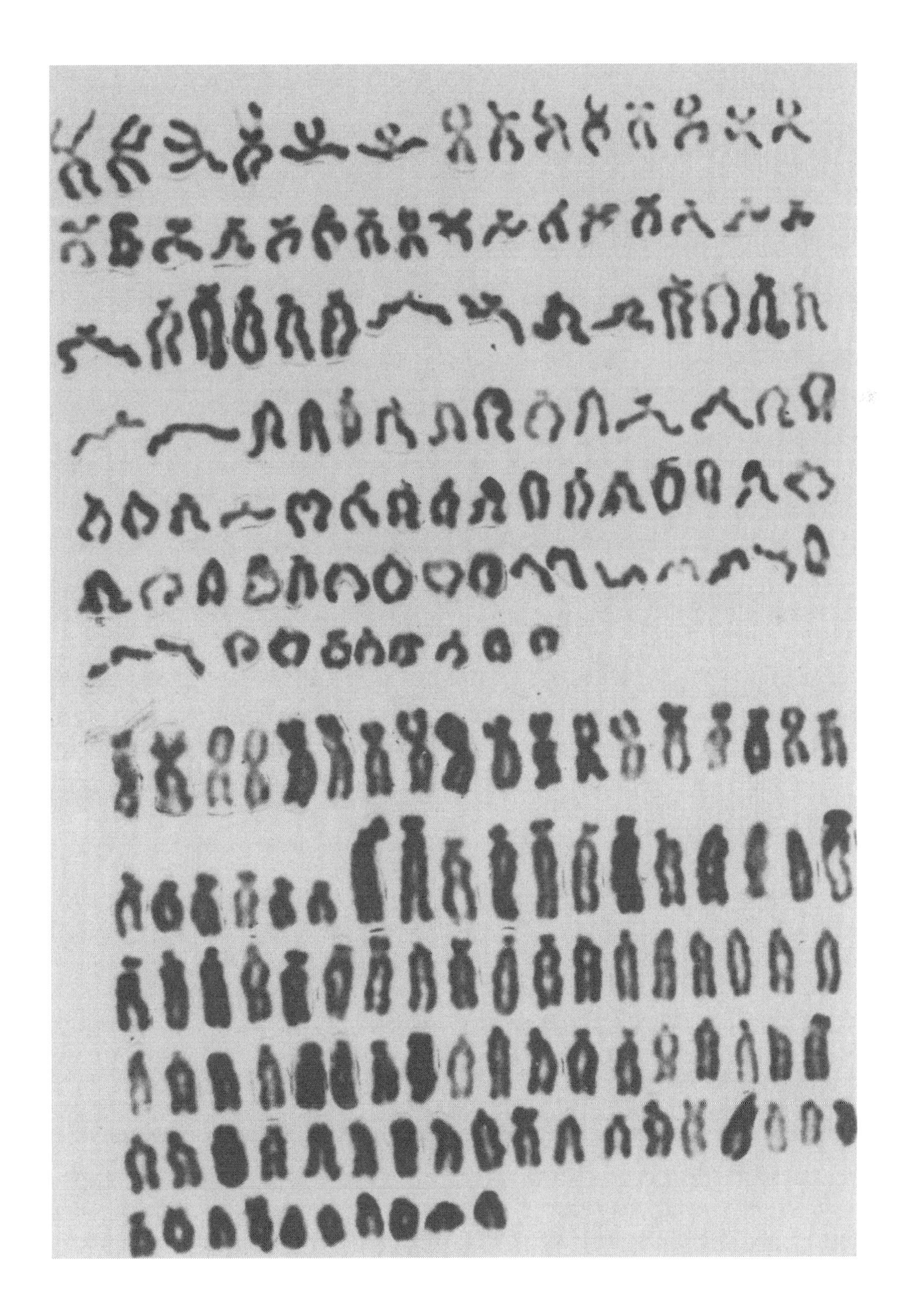

Order: ANURA

Family: BUFONIDAE

Bufo regularis

2n = 20

Volume 2, Folio Am-12, 1973

Order: ANURA Family: BUFONIDAE

Bufo regularis

2n = 20

AUTOSOMES:
8 large metacentrics and submetacentrics
4 medium-sized metacentrics and submetacentrics
4 small metacentrics
4 small submetacentrics

SEX CHROMOSOMES: No sex dimorphism

The medium-sized pair of submetacentric chromosomes contains a conspicuous secondary constriction in the long arm below the centromere.

With the exception of several species of African toads (including *Bufo regularis*), the diploid number of 22 is characteristic of the members of the genus *Bufo*. Bogart (1968) suggests that those African species which have 20 chromosomes are derived from a common ancestor possessing 22 chromosomes.

The karyotypes displayed (♂ top; ♀ bottom) were prepared by Dr. Elizabeth M. Earley, Gulf South Research Institute, New Orleans, Louisiana, USA.

Kidney cultures (♂ shown) and heart cultures (♀ shown) were used for the cytological preparations.

REFERENCES:

1) Bogart, J. P.: Chromosome number difference in the amphibian genus *Bufo*: The *Bufo regularis* species group. Evolution 22: 42-45, 1968.

2) Earley, E. M.: Karyotypes of fifteen anuran species, with particular reference to secondary constrictions. Ph.D. thesis, Tulane University, New Orleans, Louisiana, 1971.

Bufo regularis

2n = 20

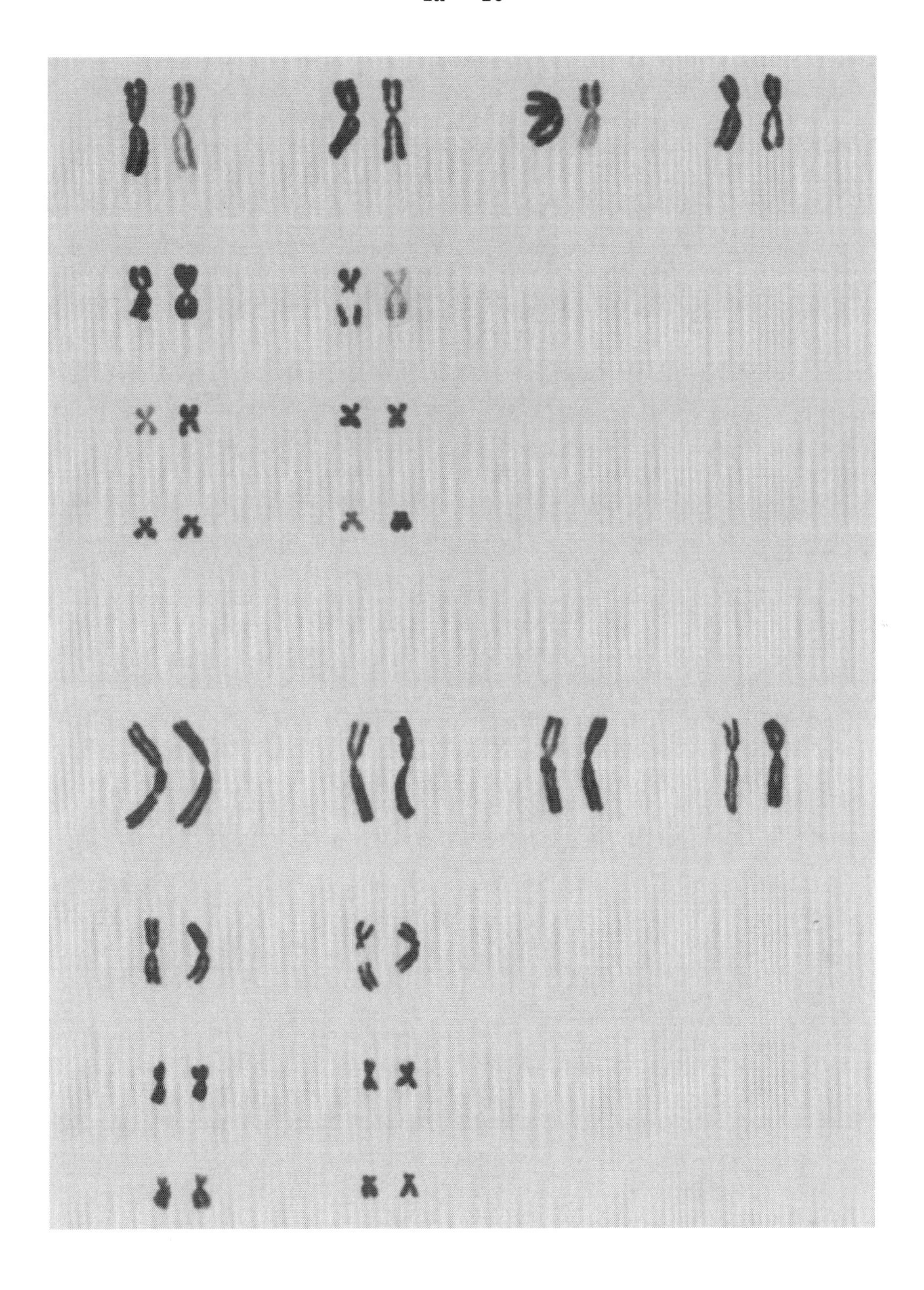

Order: ANURA

Family: BUFONIDAE

Bufo valliceps **(Wiegmann)**

(Gulf Coast toad)

2n = 22

Order: ANURA Family: BUFONIDAE

Bufo valliceps Wiegmann (Gulf Coast toad)

2n = 22

AUTOSOMES: 8 large metacentrics and submetacentrics
4 medium-sized metacentrics
10 small metacentrics and submetacentrics

SEX CHROMOSOMES: No sex dimorphism

The largest pair of chromosomes in the complement bears a prominent secondary constriction in the long arm near the centromere.

The specimens were collected in Orleans Parish, Louisiana, USA, by Mr. Dennis Duplantier, Department of Biology, Tulane University, New Orleans, Louisiana.

The karyotypes displayed (♂ top; ♀ bottom) were prepared by Dr. Elizabeth M. Earley, Gulf South Research Institute, New Orleans, Louisiana, USA.

Peripheral blood (lymphocyte) cultures were used for the cytological preparations.

REFERENCES:

1) Volpe, E. M., Duplantier, D., and Earley, E. M. Clarification of alleged "cytologically verified" hybrids between a toad and a frog. Cytogenetics 9: 161-172, 1970.

2) Earley, E. M.: Karyotypes of fifteen anuran species, with particular reference to secondary constrictions. Ph.D. thesis, Tulane University, New Orleans, Louisiana, 1971.

Bufo valliceps Wiegmann (Gulf Coast toad)

2n = 22

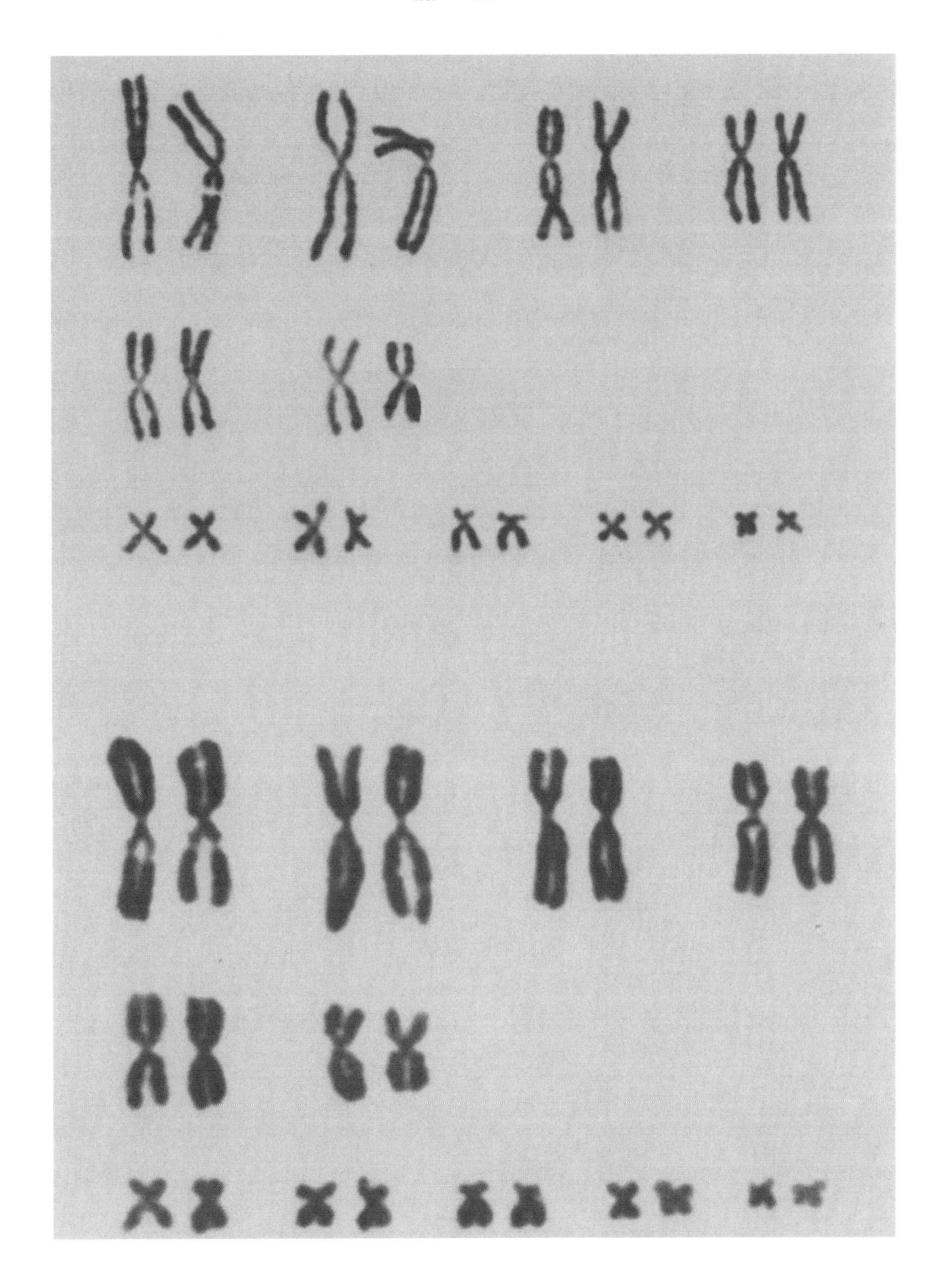

Order: ANURA

Family: HYLIDAE

Hyla chrysoscelis (Cope)
(Southern gray treefrog)
$2n = 24$

Order: ANURA Family: HYLIDAE

Hyla chrysoscelis (Cope) (Southern gray treefrog)

2n = 24

AUTOSOMES:
8 large metacentrics and submetacentrics
4 large subtelocentrics
2 medium-sized subtelocentrics
6 small submetacentrics
4 small metacentrics

SEX CHROMOSOMES: No sex dimorphism

One pair of large subtelocentric chromosomes bears a prominent secondary constriction in the short arm.

The specimens were collected in St. Tammany Parish, Louisiana, USA by Dr. Harold Dundee, Department of Biology, Tulane University, New Orleans, Louisiana.

The karyotypes displayed (♂ top; ♀ bottom) were prepared by Mr. James Turpen, Department of Biology, Tulane University, New Orleans, Louisiana, USA.

Kidney cultures (♂ shown) and muscle cultures (♀ shown) were used for the cytological preparations.

REFERENCES:

1) Bogart, J. P., and Wasserman, A. O.: Diploid-polyploid cryptic species pairs: a possible clue to evolution by polyploidization in anuran amphibians. Cytogenetics 11: 7-24, 1972.

<u>Hyla chrysoscelis</u> (Cope) (Southern gray treefrog)

2n = 24

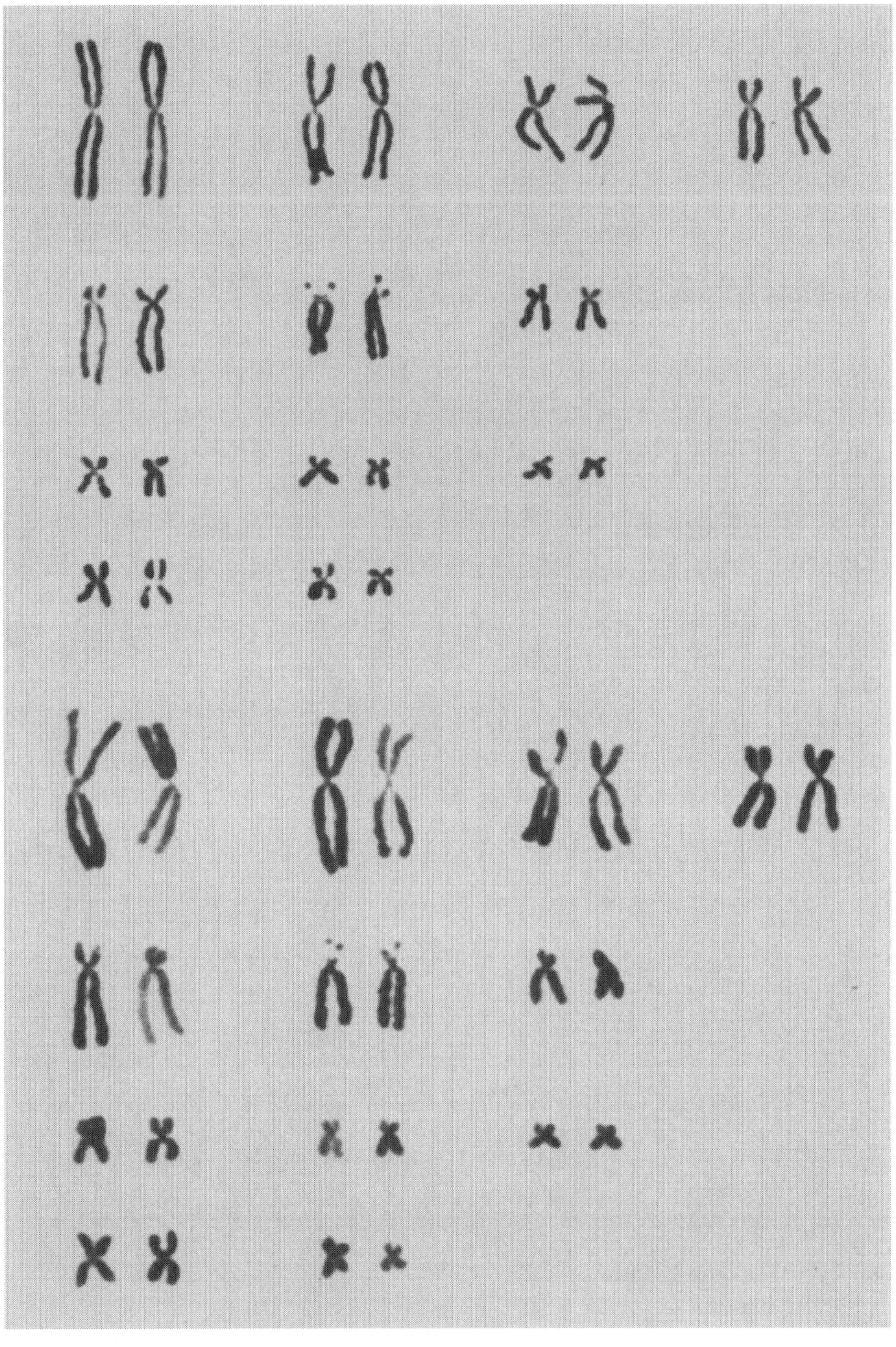

Order: ANURA

Family: HYLIDAE

Hyla cinerea (Schneider)

(Green treefrog)

2n = 24

Order: ANURA Family: HYLIDAE

Hyla cinerea (Schneider) (Green treefrog)

2n = 24

AUTOSOMES:
8 large metacentrics and submetacentrics
2 large subtelocentrics
2 medium-sized subtelocentrics
2 small subtelocentrics
6 small submetacentrics
4 small metacentrics

SEX CHROMOSOMES: No sex dimorphism

The small subtelocentric pair of chromosomes bears a prominent secondary constriction in the long arm below the centromere.

Adult specimens were collected in St. Tammany Parish, Louisiana, USA, by Dr. Harold Dundee, Department of Biology, Tulane University, New Orleans, Louisiana.

The karyotypes displayed (♂ top; ♀ bottom) were prepared by Miss Dana Reinschmidt, Department of Biology, Tulane University, New Orleans, Louisiana, USA.

Kidney cultures (♂ shown) and muscle cultures (♀ shown) were used for the cytological preparations.

Hyla cinerea (Schneider) (Green treefrog)

2n = 24

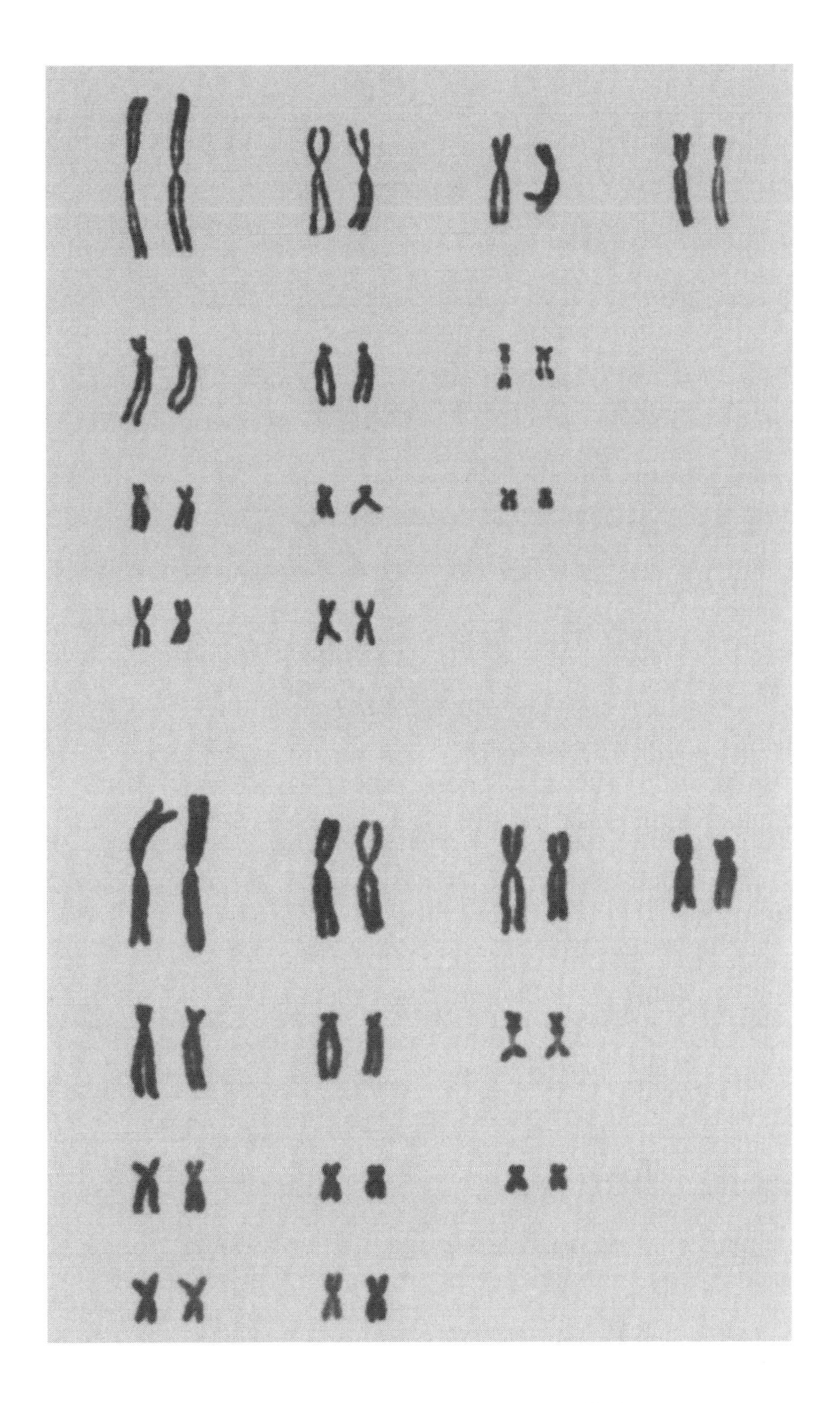

Order: ANURA

Family: HYLIDAE

Hyla versicolor (LeConte)

(Eastern gray treefrog)

$2n = 48$

Order: ANURA Family: HYLIDAE

Hyla versicolor LeConte (Eastern gray treefrog)

2n = 48

AUTOSOMES:
16 large metacentrics and submetacentrics
8 large subtelocentrics
4 medium-sized subtelocentrics
12 small submetacentrics
8 small metacentrics

SEX CHROMOSOMES: No sex dimorphism

This common treefrog represents a naturally occurring anuran polyploid in North America. The karyotype of this tetraploid species should be compared with its diploid counterpart *Hyla chrysoscelis* (Volume 2, Folio Am-14).

The karyotype of the male displayed, based on a squash preparation, was prepared by Dr. James P. Bogart of Louisiana Tech University, Ruston, Louisiana, USA. The male treefrog was derived from a population at Bastrop, Bastrop County, Texas, USA. Tetraploid treefrogs have been uncovered as well from Alpine, Bergen County, New Jersey by Dr. Aaron O. Wasserman of the City University of New York, New York, USA.

REFERENCES:

1) Wasserman, A. O.: Polyploidy in the common tree toad *Hyla versicolor* LeConte. Science 167: 385-386, 1970.

2) Bogart, J. P., and Wasserman, A. O.: Diploid-polyploid cryptic species pairs: a possible clue to evolution by polyploidization in anuran amphibians. Cytogenetics 11: 7-24, 1972.

<u>Hyla versicolor</u> LeConte (Eastern gray treefrog)

2n = 48

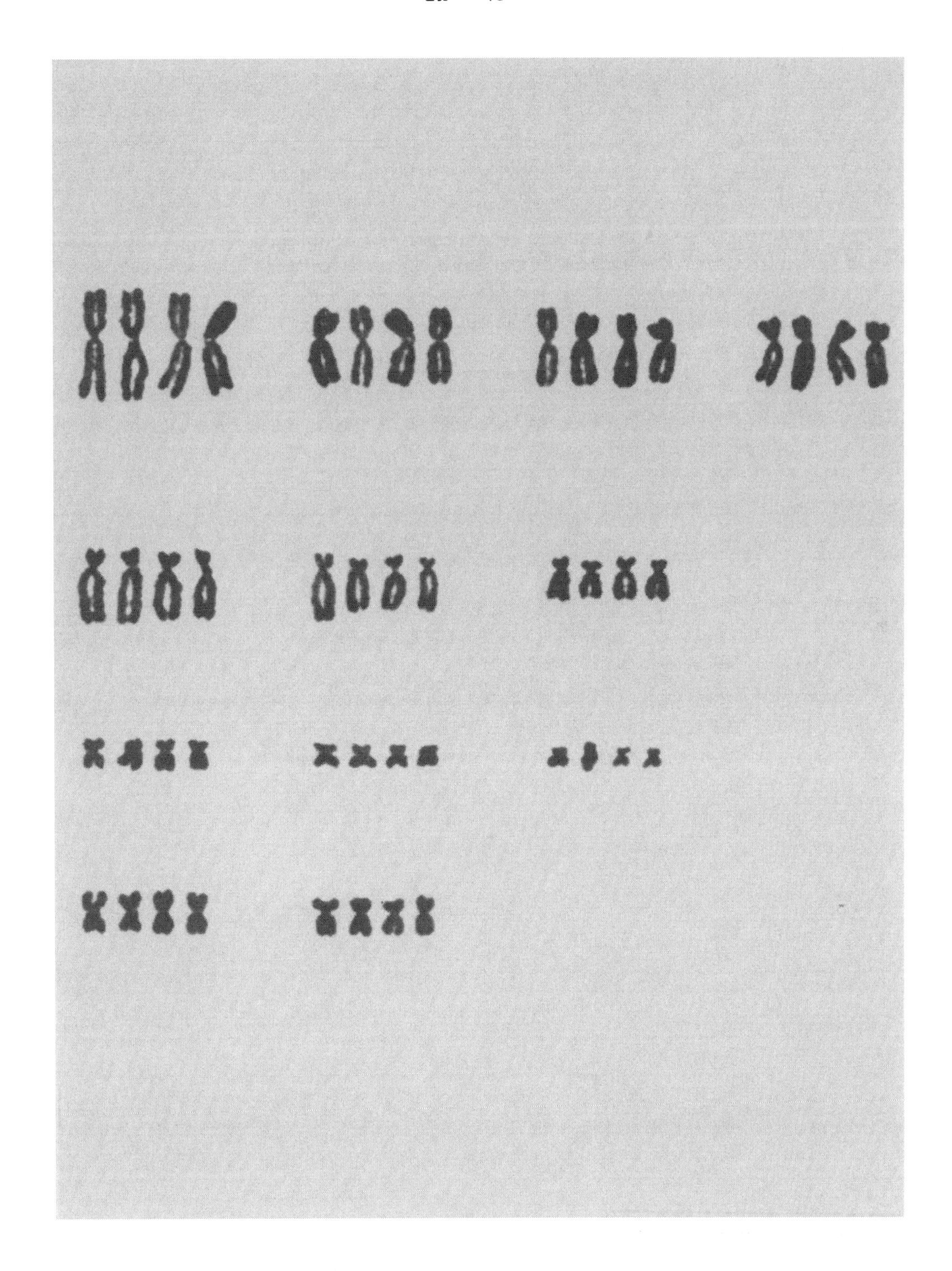

Order: ANURA

Family: PELOBATIDAE

Scaphiopus bombifrons (Cope)

(Plains spadefoot toad)

$2n = 26$

Order: ANURA　　　　Family: PELOBATIDAE

Scaphiopus bombifrons (Cope) (Plains spadefoot toad)

2n = 26

AUTOSOMES:
8 large metacentrics and submetacentrics
2 large subtelocentrics
8 small submetacentrics
8 small metacentrics

SEX CHROMOSOMES: No sex dimorphism

One pair of small submetacentric chromosomes bears a prominent secondary constriction in the long arm below the centromere.

The specimens were collected in Lubbock, Texas by Dr. Francis Rose, Department of Biology, Texas Tech University, Lubbock, Texas, USA.

The karyotypes displayed (♂ top; ♀ bottom) were prepared by Dr. Elizabeth M. Earley, Gulf South Research Institute, New Orleans, Louisiana, USA.

Karyotypes have been prepared from cultures of bone marrow, spleen, kidney, muscle, and heart tissue. Kidney cultures were used for the cytological preparations shown.

REFERENCES:

1) Wasserman, A. O., and Bogart, J. P.: Chromosomes of two species of spadefoot toads (Genus Scaphiopus) and their hybrid. Copeia, 1968: 303-306, 1968.

2) Wasserman, A. O.: Chromosomal studies of the Pelobatidae (Salientia) and some instances of ploidy. Southwest. Nat. 15: 239-248, 1970.

3) Earley, E. M.: Karyotypes of fifteen anuran species, with particular reference to secondary constrictions. Ph.D. thesis, Tulane University, New Orleans, Louisiana, 1971.

Scaphiopus bombifrons (Cope) (Plains spadefoot toad)

2n = 26

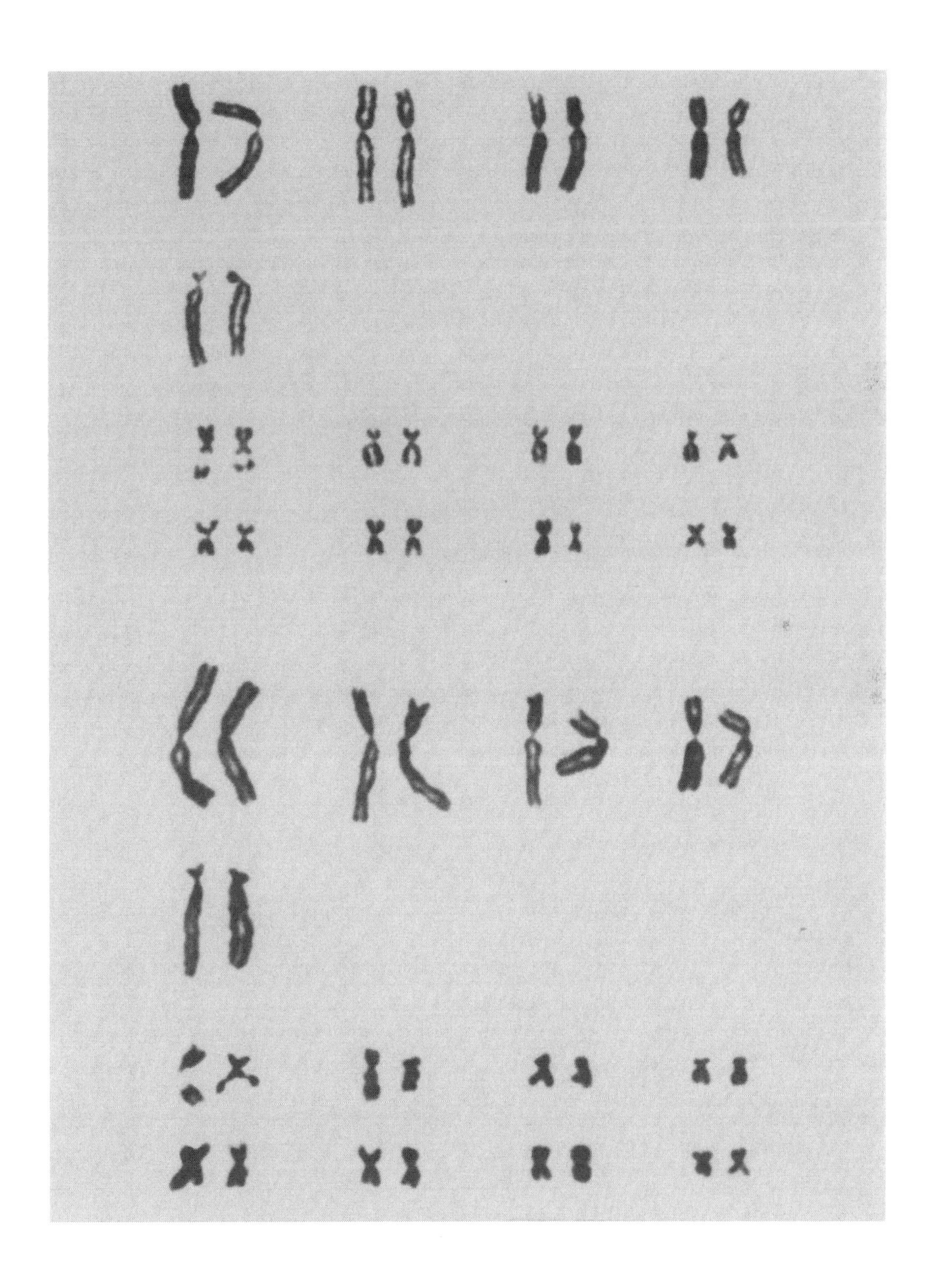

Order: ANURA

Family: PELOBATIDAE

Scaphiopus couchi (Baird)
(Couch's spadefoot toad)
$2n = 26$

Order: ANURA Family: PELOBATIDAE

Scaphiopus couchi Baird (Couch's spadefoot toad)

2n = 26

AUTOSOMES:
8 large metacentrics and submetacentrics
4 large subtelocentrics
6 small submetacentrics
8 small metacentrics

SEX CHROMOSOMES: No sex dimorphism

The smallest submetacentric chromosome bears a faint secondary constriction in the short arm above the centromere.

The specimens were collected in Morehouse Parish, Louisiana, USA, by Mr. Dennis Duplantier, Department of Biology, Tulane University, New Orleans, Louisiana.

The karyotypes displayed (♂ top; ♀ bottom) were prepared by Dr. Elizabeth M. Earley, Gulf South Research Institute, New Orleans, Louisiana, USA.

Kidney cultures (♂ shown) and bone marrow cultures (♀ shown) were used for the cytological preparations.

REFERENCES:

1) Wasserman, A. O., and Bogart, J. P.: Chromosomes of two species of spadefoot toads (Genus Scaphiopus) and their hybrid. Copeia, 1968: 303-306, 1968.

2) Wasserman, A. O.: Chromosomal studies of the Pelobatidae (Salientia) and some instances of ploidy. Southwest. Nat. 15: 239-248, 1970.

3) Earley, E. M.: Karyotypes of fifteen anuran species, with particular reference to secondary constrictions. Ph.D. thesis, Tulane University, New Orleans, Louisiana, 1971.

Scaphiopus couchi Baird (Couch's spadefoot toad)

2n = 26

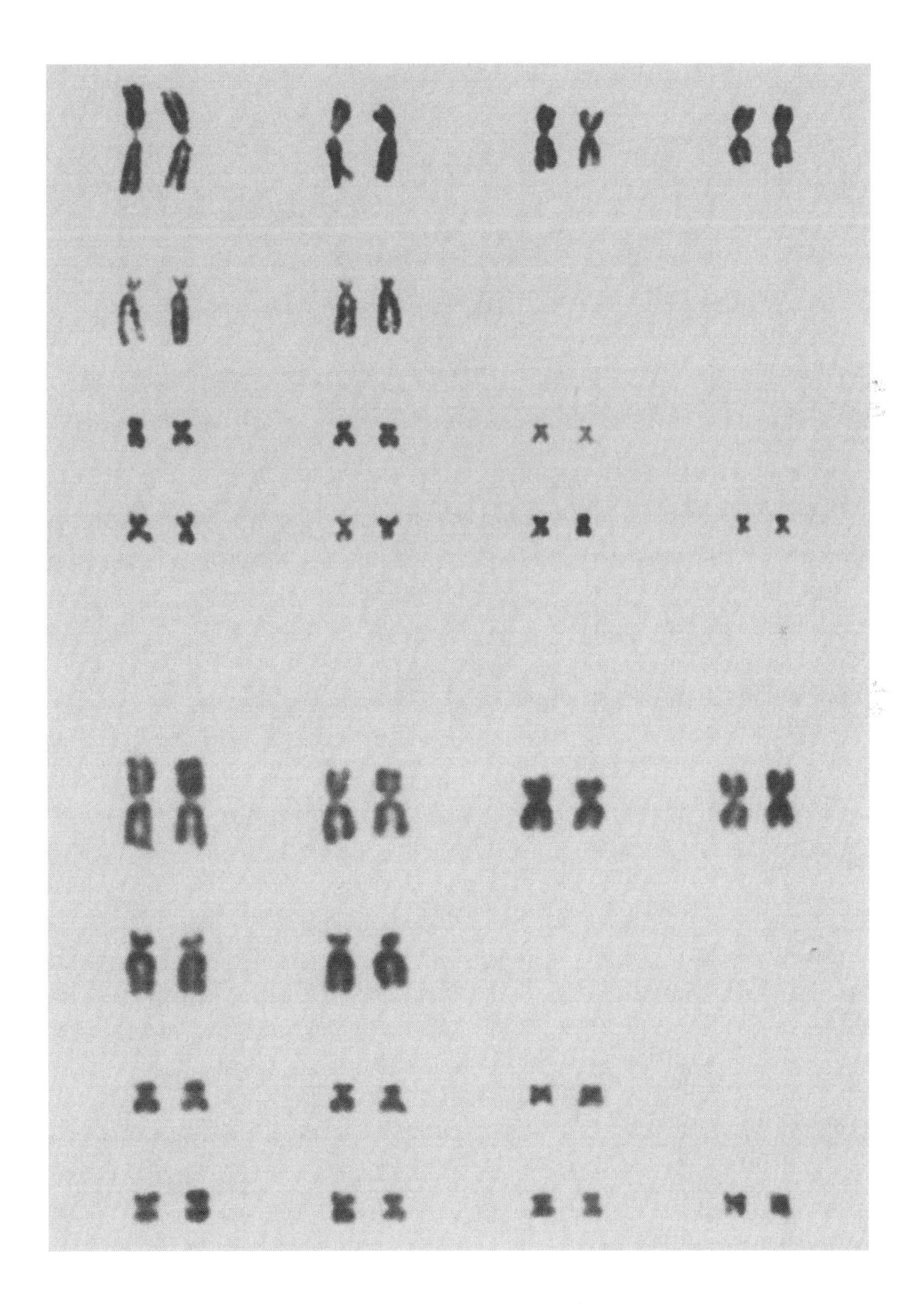

Order: ANURA

Family: RANIDAE

Rana catesbeiana (Shaw)
(Bullfrog)
$2n = 26$

Order: ANURA Family: RANIDAE

Rana catesbeiana Shaw (Bullfrog)

2n = 26

AUTOSOMES:
10 large metacentrics and submetacentrics
8 medium-sized metacentrics and submetacentrics
2 medium-sized acrocentrics (satellited)
6 small metacentrics and submetacentrics

SEX CHROMOSOMES: No sex dimorphism

One pair of medium-sized submetacentric chromosomes has a prominent secondary constriction in the long arm. Another pair of medium-sized submetacentric chromosomes has a conspicuous secondary constriction in the short arm.

Adult animals from the Wisconsin-Minnesota area of the USA were supplied by a dealer, E. G. Steinhilber & Co., Wisconsin.

The karyotypes displayed (♂ top; ♀ bottom) were prepared by Dr. Elizabeth M. Earley, Gulf South Research Institute, New Orleans, Louisiana, USA.

Lymphocyte, bone marrow, spleen, and kidney cultures were used for karyological studies. The karyotypes shown are from kidney cultures.

REFERENCES:

1) Wolf, A. H., and Quimby, M. C. Amphibian cell culture. Permanent line from the bullfrog (*Rana catesbeiana*). Science 144: 1578-1580, 1964.

2) Reynhout, J. K., and Kimmel, D. L. Chromosome studies of the lethal hybrid *Rana pipiens* ♀ X *Rana catesbeiana* ♂. Develop. Biol. 20: 501-517, 1969.

3) Earley, E. M. Karyotypes of fifteen anuran species, with particular reference to secondary constrictions. Ph.D. thesis, Tulane University, New Orleans, Louisiana, 1971.

Rana catesbeiana Shaw (Bullfrog)

2n = 26

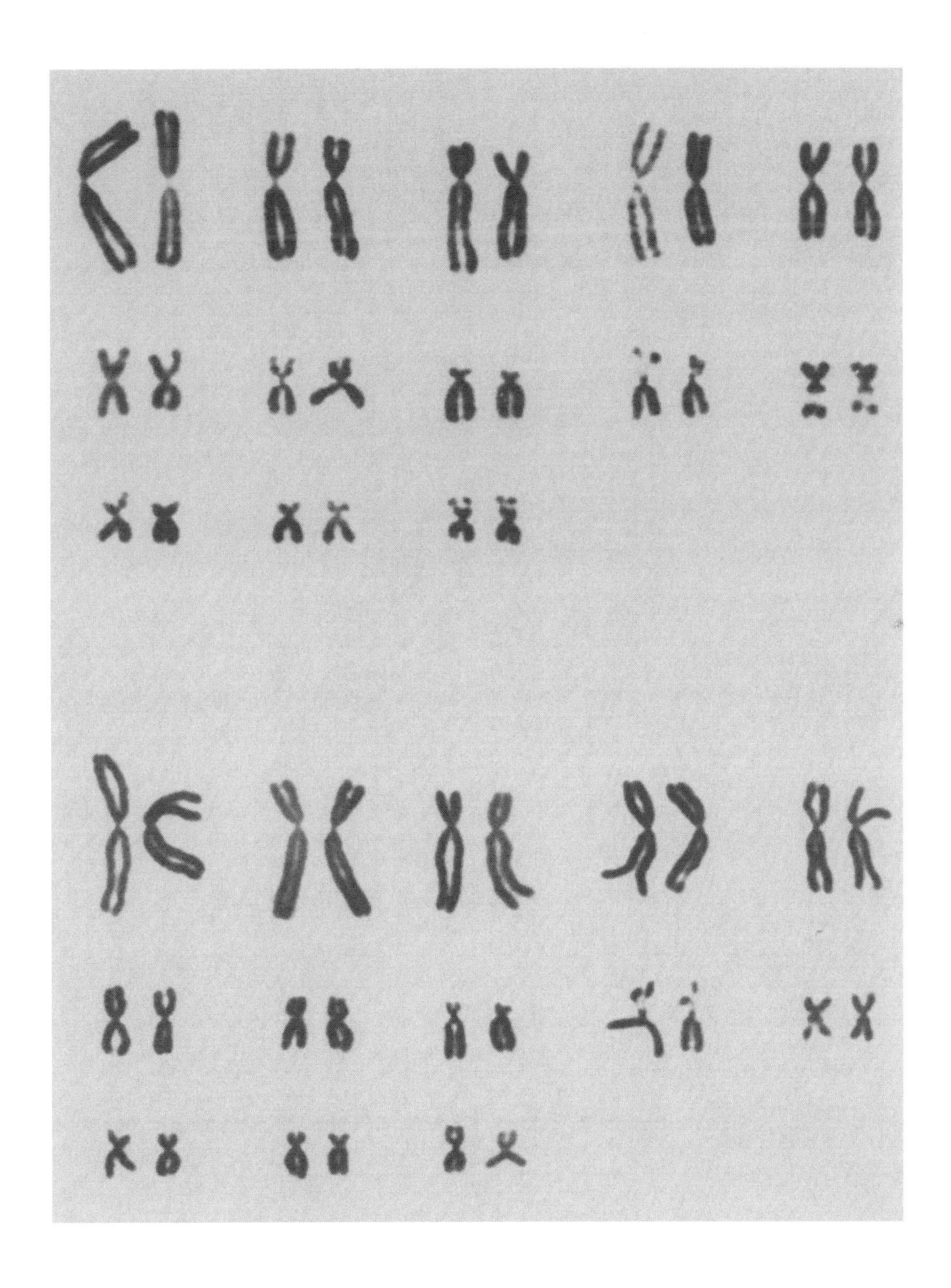

Order: ANURA

Family: RANIDAE

Rana nigromaculata (Hallowell)

(Black-spotted, or Tonosama frog)

2n = 26

Order: ANURA

Family: RANIDAE

Rana nigromaculata Hallowell (Black-spotted, or Tonosama frog)

2n = 26

AUTOSOMES:
10 large metacentrics and submetacentrics
6 medium-sized metacentrics and submetacentrics
4 medium-sized subtelocentrics
6 small metacentrics and submetacentrics

SEX CHROMOSOMES: No sex dimorphism

One pair of the smaller submetacentrics has a secondary constriction in each long arm. It is of interest that this species was originally described by Cantor in 1842 as a variety of *Rana esculenta* (see karyotype in Volume 1, Folio Am-10).

The karyotype is a gift of Dr. Midori Nishioka, Laboratory for Amphibian Biology, Faculty of Science, Hiroshima University, Hiroshima, Japan. The karyotype was analyzed from squash preparations of the tailtips of tadpoles.

REFERENCES:

1) Iriki, S.: Studies on amphibian chromosomes. IV. On the chromosomes of *Rana rugosa* and *Rana nigromaculata*. Sci. Reports of Tokyo Bunrika Daigaku, Sec. B, 1(5): 61, 1932.

2) Seto, T., and Rounds, D. E.: Cultivation of tissues and leucocytes from amphibians. *In* Methods in Cell Physiology. Academic Press, New York, 1968, pp. 75-94.

Rana nigromaculata Hallowell (Black-spotted, or Tonosama frog)

2n = 26

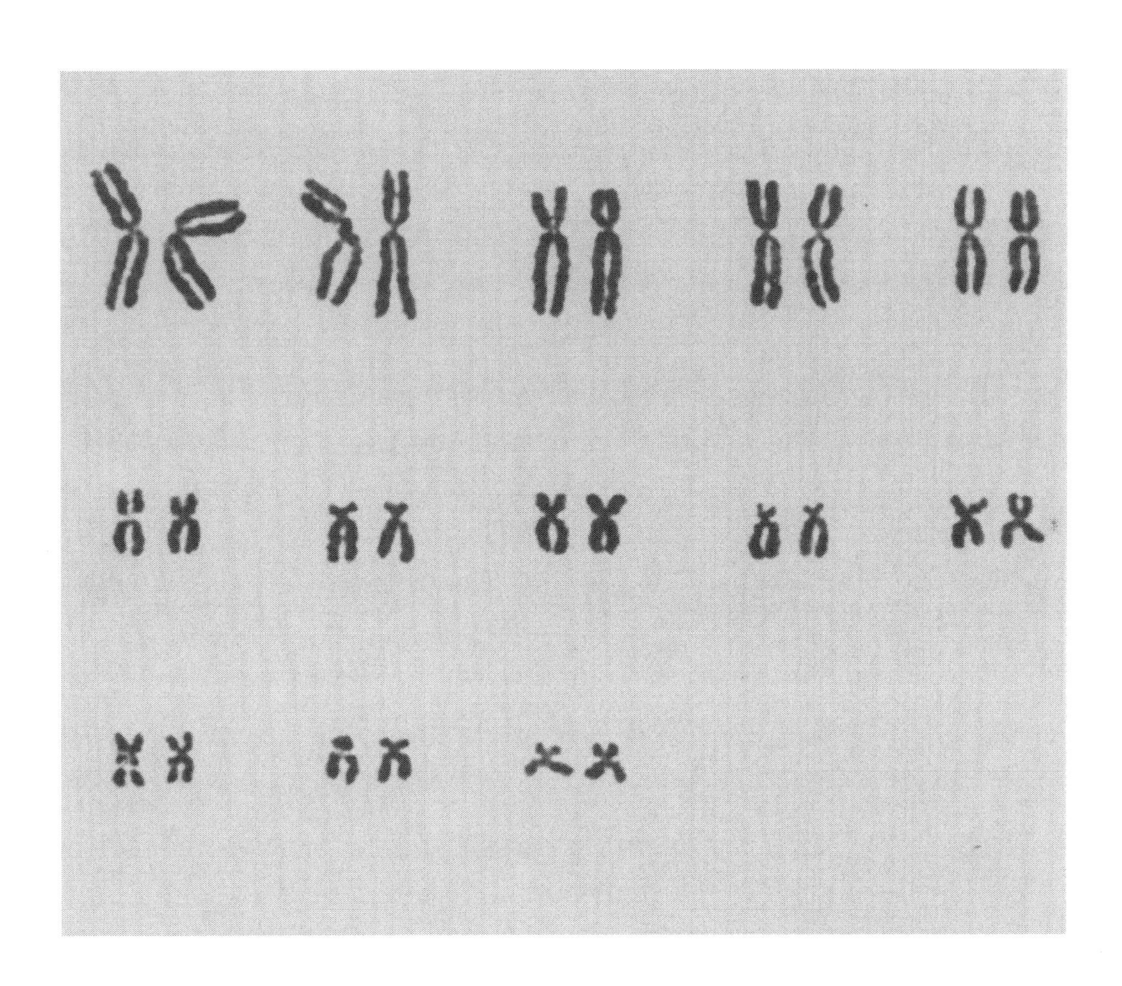

Order: ANURA

Family: RANIDAE

Rana pipiens sphenocephala (Cope)
(Southern leopard frog)
$2n = 26$

Order: ANURA Family: RANIDAE

Rana pipiens sphenocephala (Cope) (Southern leopard frog)

2n = 26

AUTOSOMES: 10 large metacentrics and submetacentrics
10 medium-sized metacentrics and submetacentrics
6 small metacentrics or submetacentrics

SEX CHROMOSOMES: No sex dimorphism

One pair of medium-sized submetacentric chromosomes (specifically, the No. 10 pair) bears a prominent secondary constriction in the long arm. The karyotype of the southern leopard frog differs from that of the northern leopard frog (see Volume 1, Folio Am-11) in that the No. 8 pair of chromosomes has a submedian centromere, rather than the median centromere that characterizes this chromosome in the northern leopard frog.

The specimens were collected in Morehouse Parish, Louisiana, USA, by Mr. Dennis Duplantier, Department of Biology, Tulane University, New Orleans, Louisiana.

The karyotypes displayed (♂ top; ♀ bottom) were prepared by Dr. Elizabeth M. Earley, Gulf South Research Institute, New Orleans, Louisiana, USA.

Kidney cultures were used for the cytological preparations.

REFERENCES:

1) Earley, E. M.: Karyotypes of fifteen anuran species, with particular reference to secondary constrictions. Ph.D. thesis, Tulane University, New Orleans, Louisiana, 1971.

Rana pipiens sphenocephala (Cope) (Southern leopard frog)

2n = 26

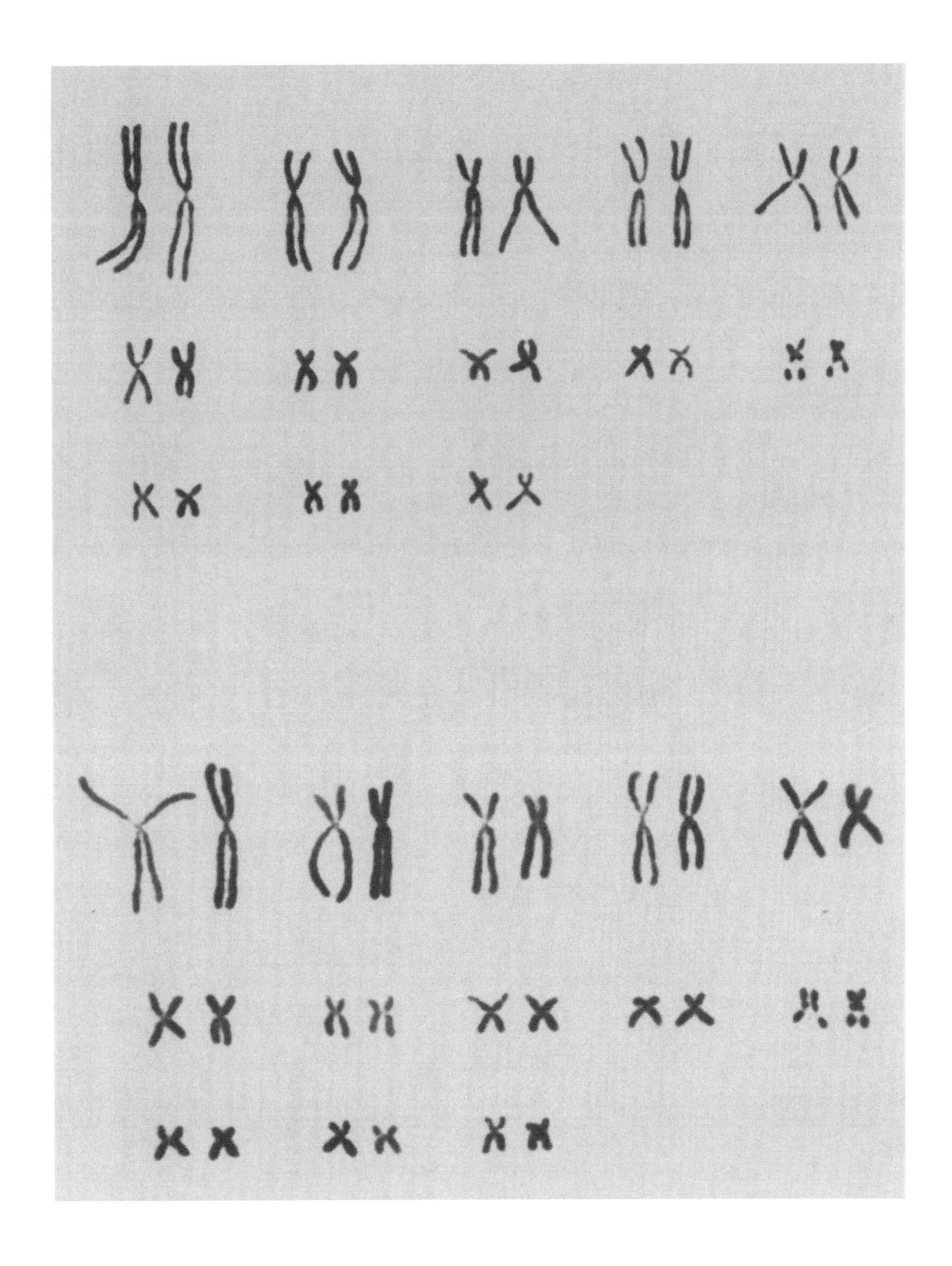

Order: ANURA

Family: RANIDAE

Rana sylvatica sylvatica (LeConte)

(Eastern wood frog)

$2n = 26$

Order: ANURA   Family: RANIDAE

Rana sylvatica sylvatica LeConte (Eastern wood frog)

2n = 26

AUTOSOMES: 10 large metacentrics and submetacentrics
8 medium-sized metacentrics and submetacentrics
2 medium-sized subtelocentrics
6 small metacentrics and submetacentrics

SEX CHROMOSOMES: No sex dimorphism

One pair of medium-sized submetacentric chromosomes bears a prominent secondary constriction in both the short and long arm. This pair of chromosomes is highly dimorphic as seen in the karyotypes displayed.

Adult animals were collected from northern Louisiana. Juveniles were laboratory-bred animals from the University of Michigan Amphibian Facility (under the direction of Dr. George Nace).

The karyotypes displayed (♂ top; ♀ bottom) were prepared by Dr. Elizabeth M. Earley, Gulf South Research Institute, New Orleans, Louisiana, USA.

Lymphocyte, bone marrow, kidney, heart, and muscle cultures were used for karyological studies. The o karyotype is from a bone marrow culture; the ♀ karyotype from a kidney culture.

REFERENCES:

1) Hennen, S.: The karyotype of *Rana sylvatica* and its comparison with the karyotype of *Rana pipiens*. J. Hered. 55: 124-128, 1964.

2) Rafferty, K. A.: Mass culture of amphibian cells: methods and observations concerning stability of cell type. *In* Biology of Amphibian Tumors (M. Mizell, ed.). Springer-Verlag, New York Inc., 1969, pp. 52-81.

Rana sylvatica sylvatica LeConte (Eastern wood frog)

2n = 26

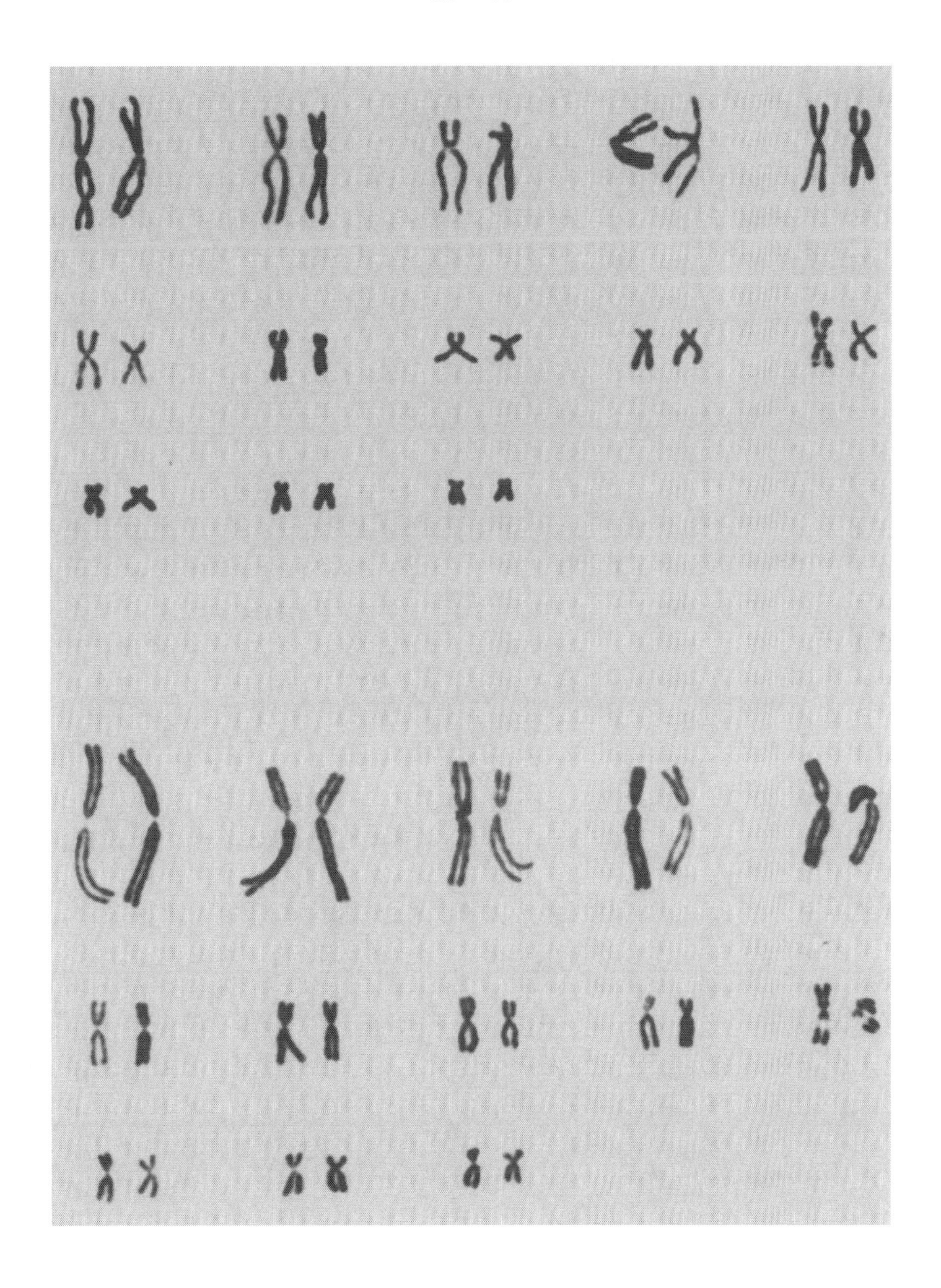

Order: URODELA

Family: SALAMANDRIDAE

Pleurodeles waltli (Michahelles)
(Spanish ribbed newt)
$2n = 24$

Order: URODELA　　　　Family: SALAMANDRIDAE

Pleurodeles waltli Michahelles (Spanish ribbed newt)

2n = 24

AUTOSOMES: 16 large metacentrics and submetacentrics
6 medium-sized submetacentrics
2 smaller subtelocentrics

SEX CHROMOSOMES: No sex dimorphism

The karyotype was derived from squash preparations of tail tips of newt larvae maintained in the Laboratoire d'Embryologie, Université de Paris, France, and provided by Professor L. Gallien, director of the culture established in the Laboratoire in 1945.

REFERENCES:

1) Gallien, C. L.: Le carotype de l'Urodèle Pleurodeles poireti Gervais. Etude comparative des caryotypes dans le genre Pleurodeles. C. R. Acad. Sc. 262: 122-125, 1966.

2) Lacroix, J. C.: Etude descriptive des chromosomes de écouvillon dans le genre Pleurodeles (Amphibian, Urodèle). Ann. Embr. Morph. 1: 179-202, 1968a.

3) Gallien, L.: Spontaneous and experimental mutations in the newt Pleurodeles waltlii Michah. In Biology of Amphibian Tumors (M. Mizell, ed.). Springer-Verlag, New York, 1969, pp. 35-42.

Order: URODELA Family: SALAMANDRIDAE

Pleurodeles waltli Michahelles (Spanish ribbed newt)

2n = 24

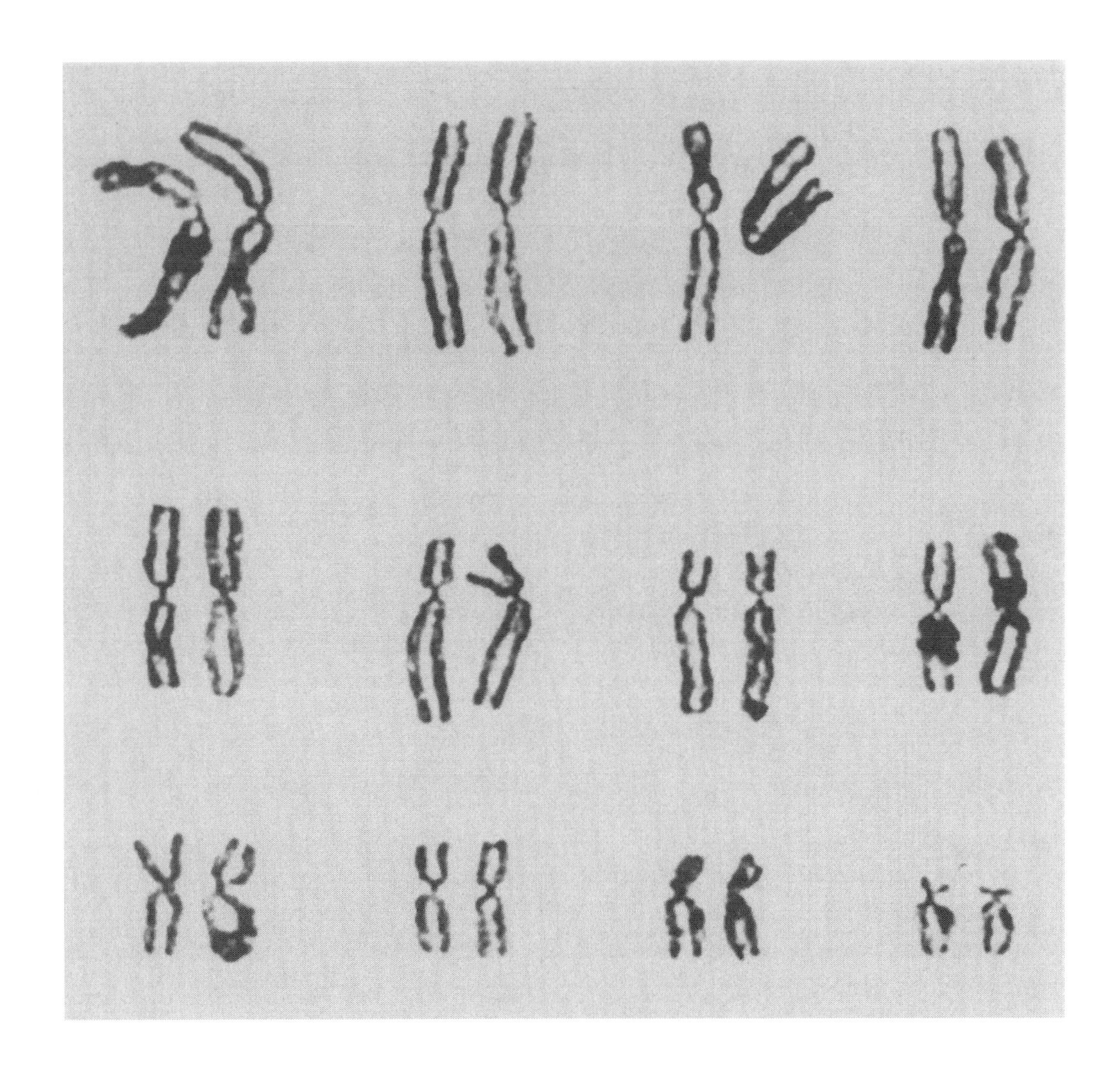

Order: SQUAMATA
Suborder: SERPENTES

Family: VIPERIDAE

Bothrops alternatus (Duméril, Bibron et Duméril)

(Urutu cruzeiro; Jararaca rabo de porco)

2n = 36

Order: SQUAMATA Family: VIPERIDAE
Suborder: SERPENTES

Bothrops alternatus Duméril, Bibron et Duméril

(Urutu cruzeiro; Jararaca rabo de porco)

2n = 36

MACROCHROMOSOMES: AUTOSOMES: 10 Metacentrics and submetacentrics
4 Acrocentrics

SEX CHROMOSOMES: Z Metacentric
W Acrocentric

MICROCHROMOSOMES: 20

Both karyotypes presented here were obtained from squash preparations of intestine. The female is heterogametic. The sex chromosomes, male-ZZ, female-ZW correspond to the 4th pair of macrochromosomes.

Both specimens were captured in Paraná, Brazil. They are preserved in the Collection of the Instituto Butantan.

REFERENCES:

1) Beçak, W.: Constituição cromossômica e mecanismo de determinação do sexo em ofídios sul-americanos. I. Aspectos cariotípicos. Mem. Inst. Butantan 32: 37, 1965.

2) Beçak, W.: Constituição cromossômica e mecanismo de determinação do sexo em ofídios sul-americanos. II. Cromossomos sexuais e evolução do cariótipo. Mem. Inst. Butantan. Simp. int. 33: 775, 1966.

3) Beçak, W.: Karyotypes, sex chromosomes and chromosomal evolution in snakes. In "Venomous Animals and their Venoms" (W. Bücherl, E. Buckley and V. Deulofeu, eds.) Vol. I, p. 53. Academic Press,New York 1968.

4) Beçak, W. and Beçak, M. L.: Cytotoxonomy and chromosomal evolution in Serpentes. Cytogenetics 8: 247, 1969.

Order: SQUAMATA
Suborder: SERPENTES

Family: VIPERIDAE

Bothrops alternatus Duméril, Bibron et Duméril

(Urutu cruzeiro; Jararaca rabo de porco)

2n = 36

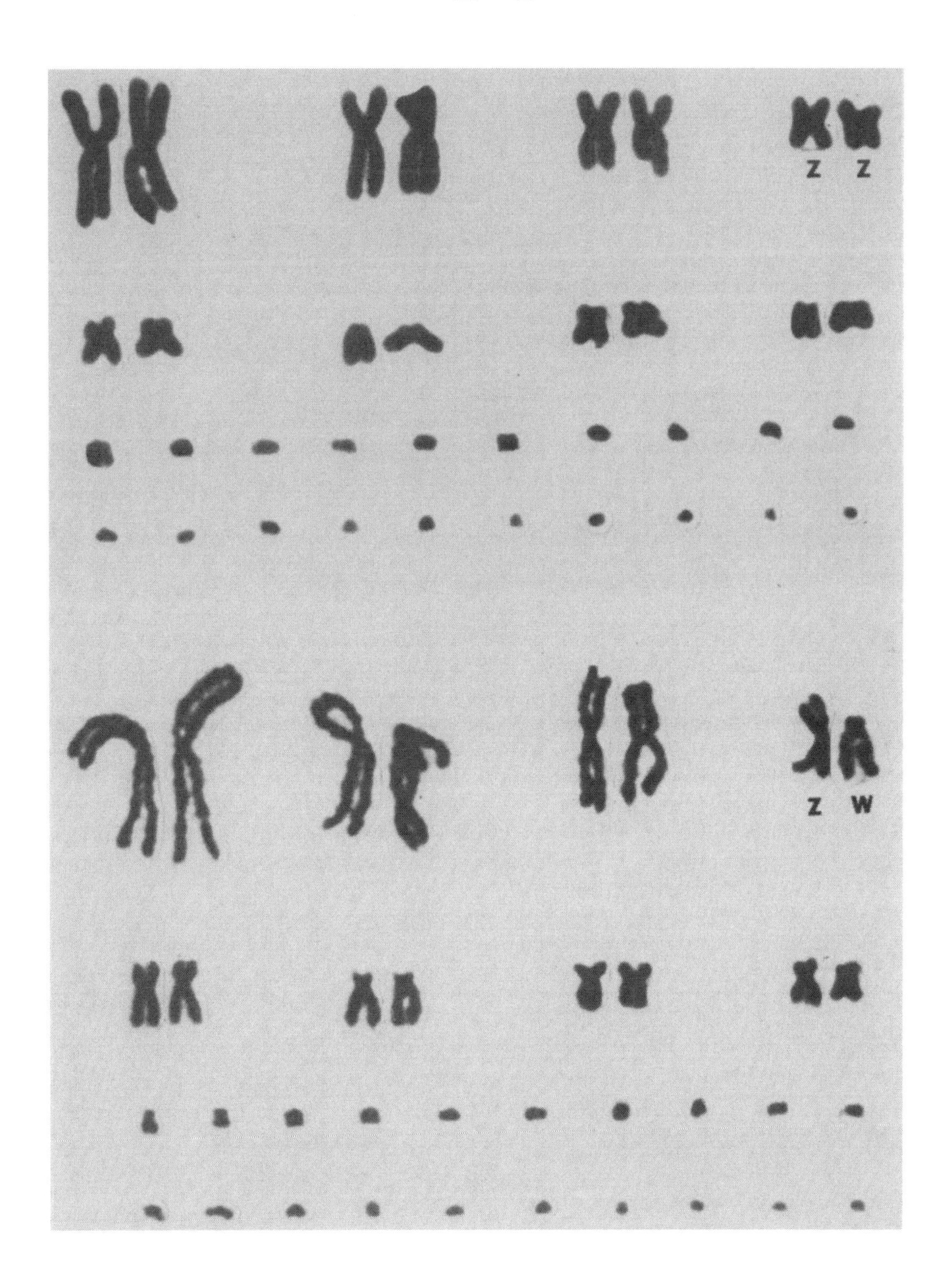

Order: SQUAMATA
Suborder: SERPENTES

Family: COLUBRIDAE

Clelia occipitolutea (Duméril, Bibron et Duméril)

(Black snake Muçurana; Cobra preta)

2n = 50

Order: SQUAMATA
Suborder: SERPENTES

Family: COLUBRIDAE

Clelia occipitolutea (Duméril, Bibron et Duméril)

(Black snake Muçurana; Cobra preta)

2n = 50

MACROCHROMOSOMES: AUTOSOMES: 12 Acrocentrics

SEX CHROMOSOMES: Z Metacentric
W Acrocentric

MICROCHROMOSOMES: 36

The karyotypes presented here were obtained from short term blood cultures. The sex chromosomes, male-ZZ, female-ZW correspond to the 4th pair of macrochromosomes. The W is about twice as large as the Z. The diploid number of 50 chromosomes is the highest so far reported in snakes.

Both specimens were captured in Brazil, the male in Rio Grande do Sul. Both are preserved in the Collection of the Instituto Butantan.

REFERENCES:

1) Beçak, W., Beçak, M. L., Nazareth, H. R. S. and Ohno, S.: Close karyological kinship between the reptilian suborder Serpentes and the class Aves. Chromosoma (Berl.) 15: 606, 1964.

2) Beçak, W.: Constituição cromossômica e mecanismo de determinação do sexo em ofídios sul-americanos. I. Aspectos cariotípicos. Mem. Inst. Butantan 32: 37, 1965.

3) Beçak, W.: Constituição cromossômica e mecanismo de determinação do sexo em ofídios sul-americanos. II. Cromossomos sexuais e evolução do cariótipo. Mem. Inst. Butantan. Simp. int. 33: 775, 1966.

4) Beçak, W.: Karyotypes, sex chromosomes and chromosomal evolution in snakes. In "Venomous Animals and their Venoms" (W. Bücherl, E. Buckley and V. Deulofeu, eds.) Vol. I, p. 53. Academic Press,New York 1968.

5) Beçak, W. and Beçak, M. L.: Cytotaxonomy and chromosomal evolution in Serpentes. Cytogenetics 8: 247, 1969.

Order: SQUAMATA
Suborder: SERPENTES

Family: COLUBRIDAE

<u>Clelia occipitolutea</u> (Duméril, Bibron et Duméril)

(Black snake Muçurana; Cobra preta)

2n = 50

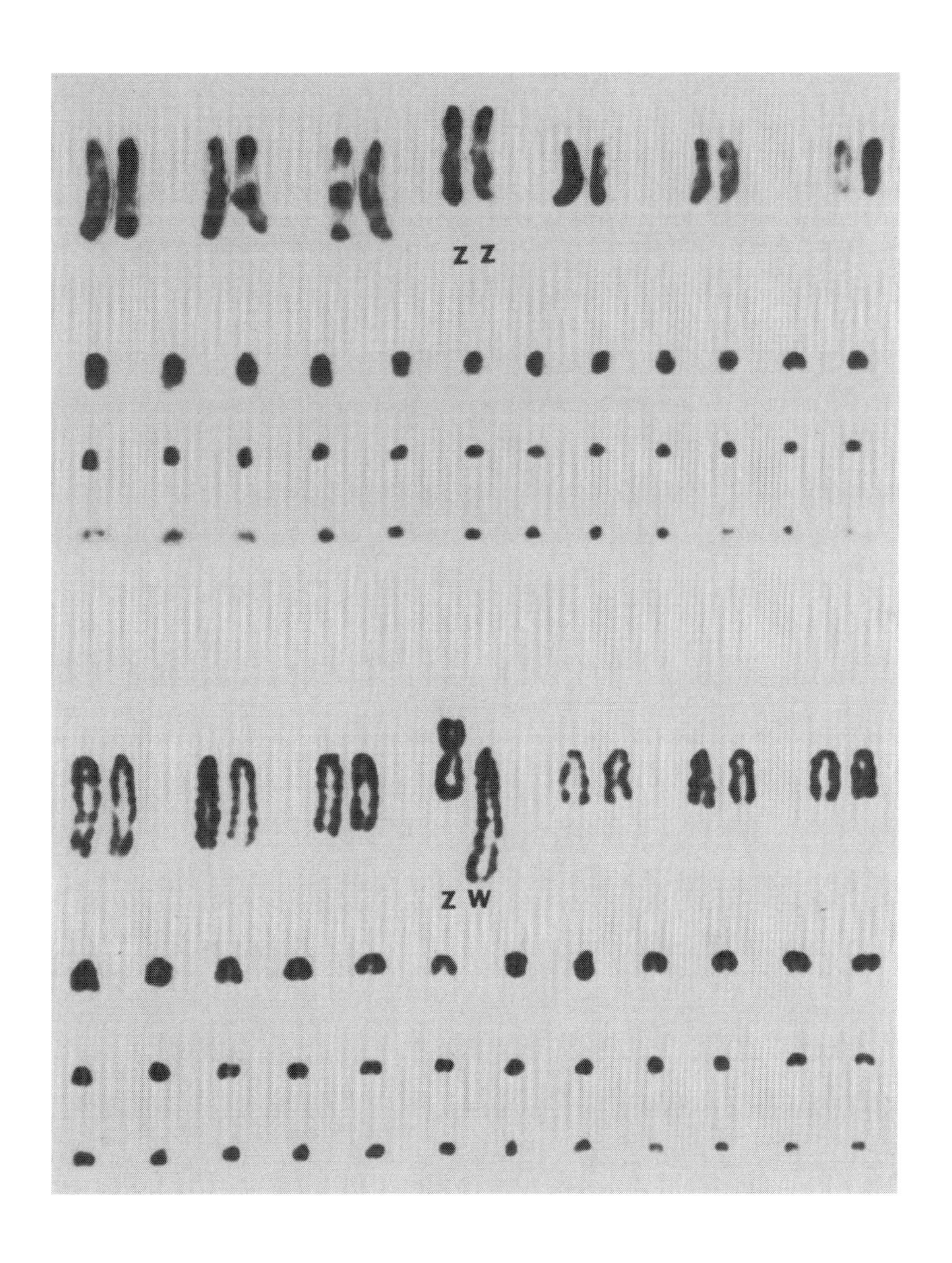

Order: SQUAMATA
Suborder: SERPENTES

Family: COLUBRIDAE

Drymarchon corais corais (Boie)

(South American indigo snake; Papa-ovo; Papa-pinto)

2n = 36

Order: SQUAMATA Family: COLUBRIDAE
Suborder: SERPENTES

Drymarchon corais corais (Boie)

(South American indigo snake; Papa-ovo; Papa-pinto)

2n = 36

MACROCHROMOSOMES: AUTOSOMES: 12 Metacentrics and submetacentrics
2 Acrocentrics

SEX CHROMOSOMES: Z Submetacentric
W Acrocentric

MICROCHROMOSOMES: 14

The karyotypes presented here were obtained from short term blood cultures. The female is heterogametic. The sex chromosomes, male-ZZ, female-ZW correspond to the 4th pair of macrochromosomes.

The karyotypes are from a male specimen of São Paulo and female specimen of Bahia, Brazil. Both are included in the Collection of the Instituto Butantan.

REFERENCES:

1) Beçak, W.: Constituição cromossômica e mecanismo de determinação do sexo em ofídios sul-americanos. I. Aspectos cariotípicos. Mem. Inst. Butantan 32: 37, 1965.

2) Beçak, W.: Constituição cromossômica e mecanismo de determinação do sexo em ofídios sul-americanos. II. Cromossomos sexuais e evolução do cariótipo. Mem. Inst. Butantan. Simp. int. 33: 775, 1966.

3) Beçak, W.: Karyotypes, sex chromosomes and chromosomal evolution in snakes. In "Venomous Animals and their Venoms" (W. Bücherl, E. Buckley and V. Deulofeu, eds.) Vol. I, p. 53. Academic Press,New York 1968.

4) Beçak, W. and Beçak, M. L.: Cytotaxonomy and chromosomal evolution in Serpentes. Cytogenetics 8: 247, 1969.

Order: SQUAMATA
Suborder: SERPENTES

Family: COLUBRIDAE

<u>Drymarchon</u> <u>corais</u> <u>corais</u> (Boie)

(South American indigo snake; Papa-ovo; Papa-pinto)

2n = 36

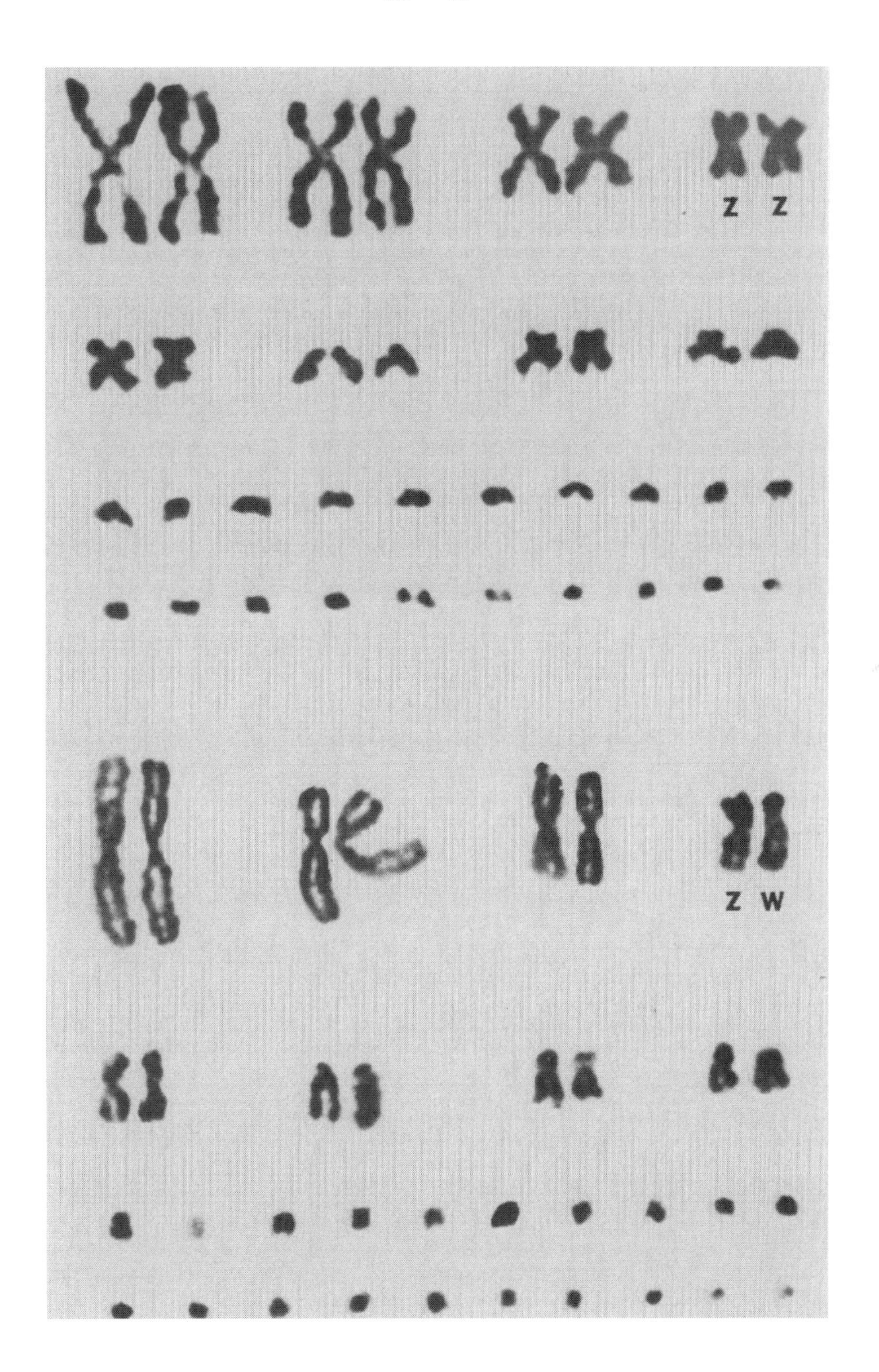

Order: SQUAMATA
Suborder: SERPENTES

Family: COLUBRIDAE

Liophis miliaris (Linnaeus)

(Cobra d'agua)

2n = 28

Order: SQUAMATA
Suborder: SERPENTES

Family: COLUBRIDAE

Liophis miliaris (Linnaeus)

(Cobra d'agua)

2n = 28

MACROCHROMOSOMES: AUTOSOMES: 26 Metacentrics and submetacentrics

SEX CHROMOSOMES: Z Metacentric
W Acrocentric

The karyotypes were obtained from squash preparations of intestine (male) and gonads (female). The female is heterogametic. The sex chromosomes, male-ZZ, female-ZW correspond to the 4th pair of macrochromosomes.

Both snakes were captured in Brazil, the male specimen in Santa Catarina. Both are preserved in the Collection of the Instituto Butantan.

REFERENCES:

1) Beçak, W.: Karyotypes, sex chromosomes and chromosomal evolution in snakes. In "Venomous Animals and their Venoms" (W. Bücherl, E. Buckley and V. Deulofeu, eds.) Vol. I, p. 53. Academic Press,New York 1968.

Order: SQUAMATA
Suborder: SERPENTES

Family: COLUBRIDAE

Liophis miliaris (Linnaeus)

(Cobra d'agua)

2n = 28

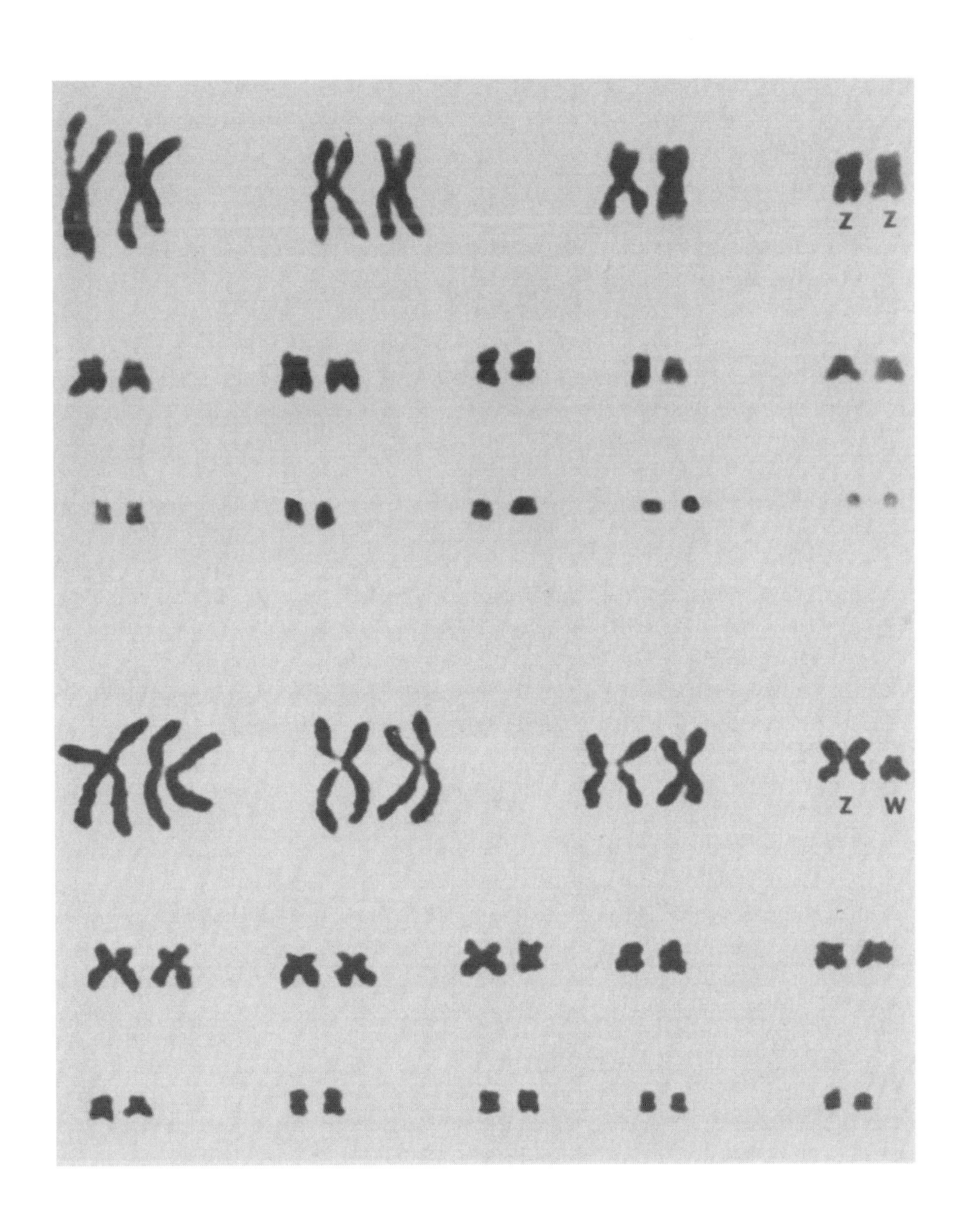

Order: SQUAMATA
Suborder: SERPENTES

Family: COLUBRIDAE

Mastigodryas bifossatus bifossatus (Raddi)

(Panther snake; Jararacuçu do brejo; Nyakaniná)

2n = 36

Order: SQUAMATA
Suborder: SERPENTES

Family: COLUBRIDAE

Mastigodryas bifossatus bifossatus (Raddi)

(Panther snake; Jararacuçu do brejo; Nyakaniná)

2n = 36

MACROCHROMOSOMES: AUTOSOMES: 14 Metacentrics and submetacentrics

SEX CHROMOSOMES: Z Metacentric
W Acrocentric

MICROCHROMOSOMES: 20

Short term cultures of blood furnished the karyotypes here presented. The 4th pair of macrochromosomes is heteromorphic in the female (ZW) and homomorphic in the male (ZZ).

The karyotypes are from a male and female specimen collected in Minas Gerais, Brazil. Both are preserved in the Collection of the Instituto Butantan.

The karyotype of this species was originally reported[1] under its former name Dryadophis bifossatus bifossatus.

REFERENCES:

1) Beçak, W.: Constituição cromossômica e mecanismo de determinação do sexo em ofídios sul-americanos. I. Aspectos cariotípicos. Mem. Inst. Butantan 32: 37, 1965.

2) Beçak, W.: Constituição cromossômica e mecanismo de determinação do sexo em ofídios sul-americanos. II. Cromossomos sexuais e evolução do cariótipo. Mem. Inst. Butantan. Simp. int. 33: 775, 1966.

3) Beçak, W.: Karyotypes, sex chromosomes and chromosomal evolution in snakes. In "Venomous Animals and their Venoms" (W. Bücherl, E. Buckley and V. Deulofeu, eds.) Vol. I, p. 53. Academic Press,New York 1968.

4) Beçak, W. and Beçak, M. L.: Cytotaxonomy and chromosomal evolution in Serpentes. Cytogenetics 8: 247, 1969.

Mastigodryas bifossatus bifossatus (Raddi)

(Panther snake; Jararacuçu do brejo; Nyakaniná)

2n = 36

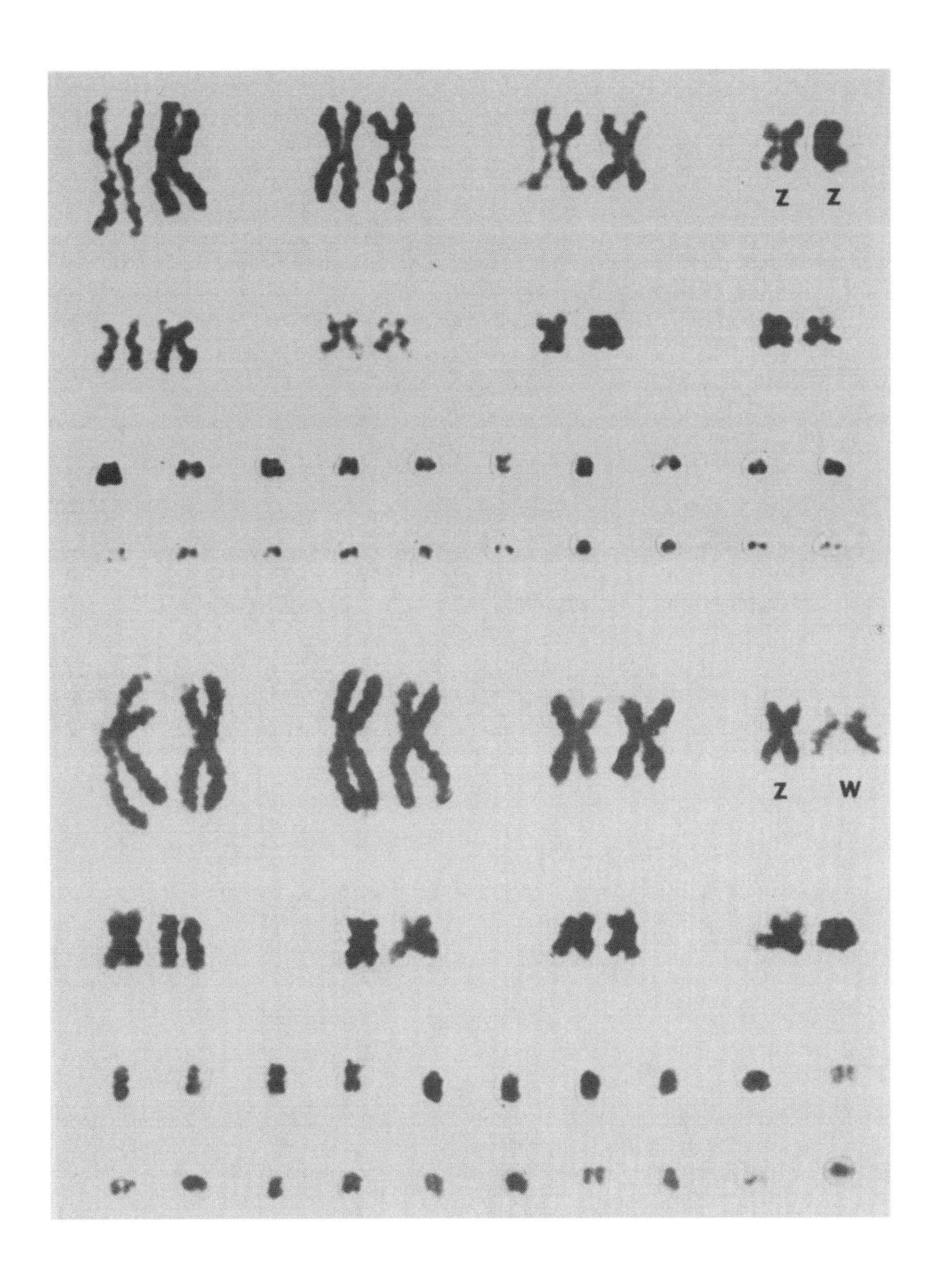

Order: SQUAMATA
Suborder: SERPENTES

Family: COLUBRIDAE

Philodryas serra (Schlegel)

(Cobra cipó)

$2n = 28$

Order: SQUAMATA
Suborder: SERPENTES

Family: COLUBRIDAE

<u>Philodryas serra</u> (Schlegel)

(Cobra cipó)

2n = 28

MACROCHROMOSOMES: AUTOSOMES: 22 Metacentrics and submetacentrics
4 Acrocentrics

SEX CHROMOSOMES: Z Metacentric
W Acrocentric

The karyotypes shown came from short term blood culture (female) and squash preparation of intestine (male). The female is heterogametic, the fourth pair of macrochromosomes consists of the sex chromosomes ZZ in the male and ZW in the female.

The male karyotype was obtained from a specimen captured in Espirito Santo and the female one from a specimen from Minas Gerais, Brazil. Both are included in the Collection of the Instituto Butantan.

REFERENCES:

1) Beçak, W., Beçak, M. L. and Nazareth, H. R. S.: Evolution and sex chromosomes in Serpentes. Mem. Inst. Butantan, Simp. int. <u>33</u>: 151, 1966.

2) Beçak, W. and Beçak, M. L.: Cytotaxonomy and chromosomal evolution in Serpentes. Cytogenetics <u>8</u>: 247, 1969.

Order: SQUAMATA
Suborder: SERPENTES

Family: COLUBRIDAE

Philodryas serra (Schlegel)

(Cobra cipó)

2n = 28

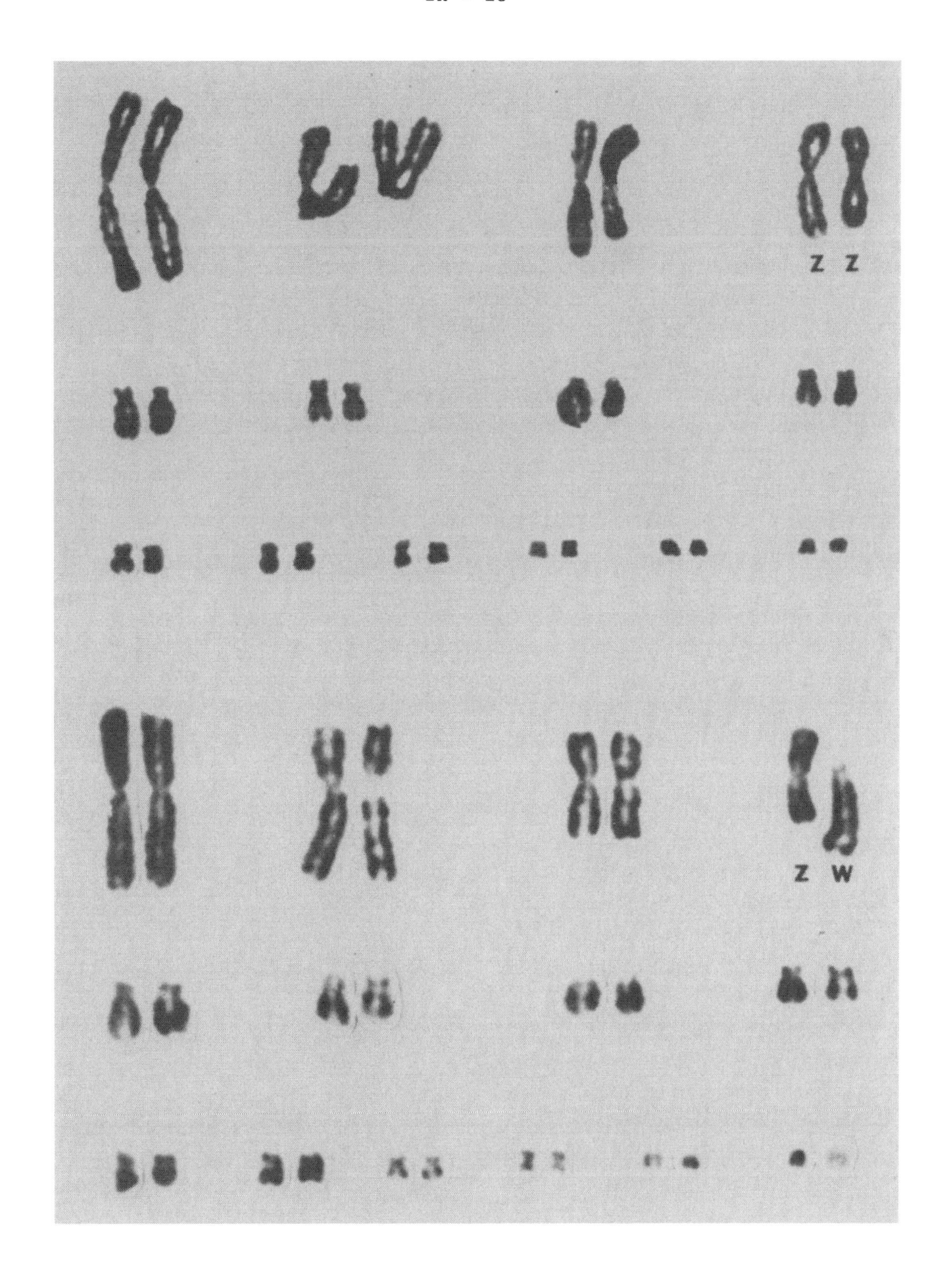

Order: SQUAMATA
Suborder: SERPENTES

Family: COLUBRIDAE

Spilotes pullatus anomalepis (Bocourt)

(Chicken snake; Caninana; Caninana do papo amarelo; Yacaniña)

2n = 36

Order: SQUAMATA
Suborder: SERPENTES

Family: COLUBRIDAE

Spilotes pullatus anomalepis Bocourt

(Chicken snake; Caninana; Caninana do papo amarelo; Yacaniña)

2n = 36

MACROCHROMOSOMES: AUTOSOMES: 12 Metacentrics and submetacentrics
2 Acrocentrics

SEX CHROMOSOMES: Z Submetacentric
W Acrocentric

MICROCHROMOSOMES: 20

The karyotypes shown came from short term blood cultures. The female is heterogametic, the fourth pair of macrochromosomes consists of the sex chromosomes ZZ in the male, and ZW in the female.

The karyotypes are from specimens collected in Paraná, Brazil. Both are preserved in the Collection of the Instituto Butantan.

REFERENCES:

1) Beçak, W., Beçak, M. L., Nazareth, H. R. S. and Ohno, S.: Close karyological kinship between the reptilian suborder Serpentes and the class Aves. Chromosoma (Berl.) 15: 606, 1964.

2) Beçak, W.: Constituição cromossômica e mecanismo de determinação do sexo em ofídios sul-americanos. I. Aspectos cariotípicos. Mem. Inst. Butantan 32: 37, 1965.

3) Beçak, W.: Constituição cromossômica e mecanismo de determinação do sexo em ofídios sul-americanos. II. Cromossomos sexuais e evolução do cariótipo. Mem. Inst. Butantan, Simp. int. 33: 775, 1966.

4) Beçak, W.: Karyotypes, sex chromosomes and chromosomal evolution in snakes. In "Venomous Animals and their Venoms" (W. Bücherl, E. Buckley and V. Deulofeu, eds.) Vol. I, p. 53. Academic Press,New York 1968.

5) Beçak, W. and Beçak, M. L.: Cytotaxonomy and chromosomal evolution in Serpentes. Cytogenetics 8: 247, 1969.

Spilotes pullatus anomalepis Bocourt

(Chicken snake; Caninana; Caninana do papo amarelo; Yacaniña)

2n = 36

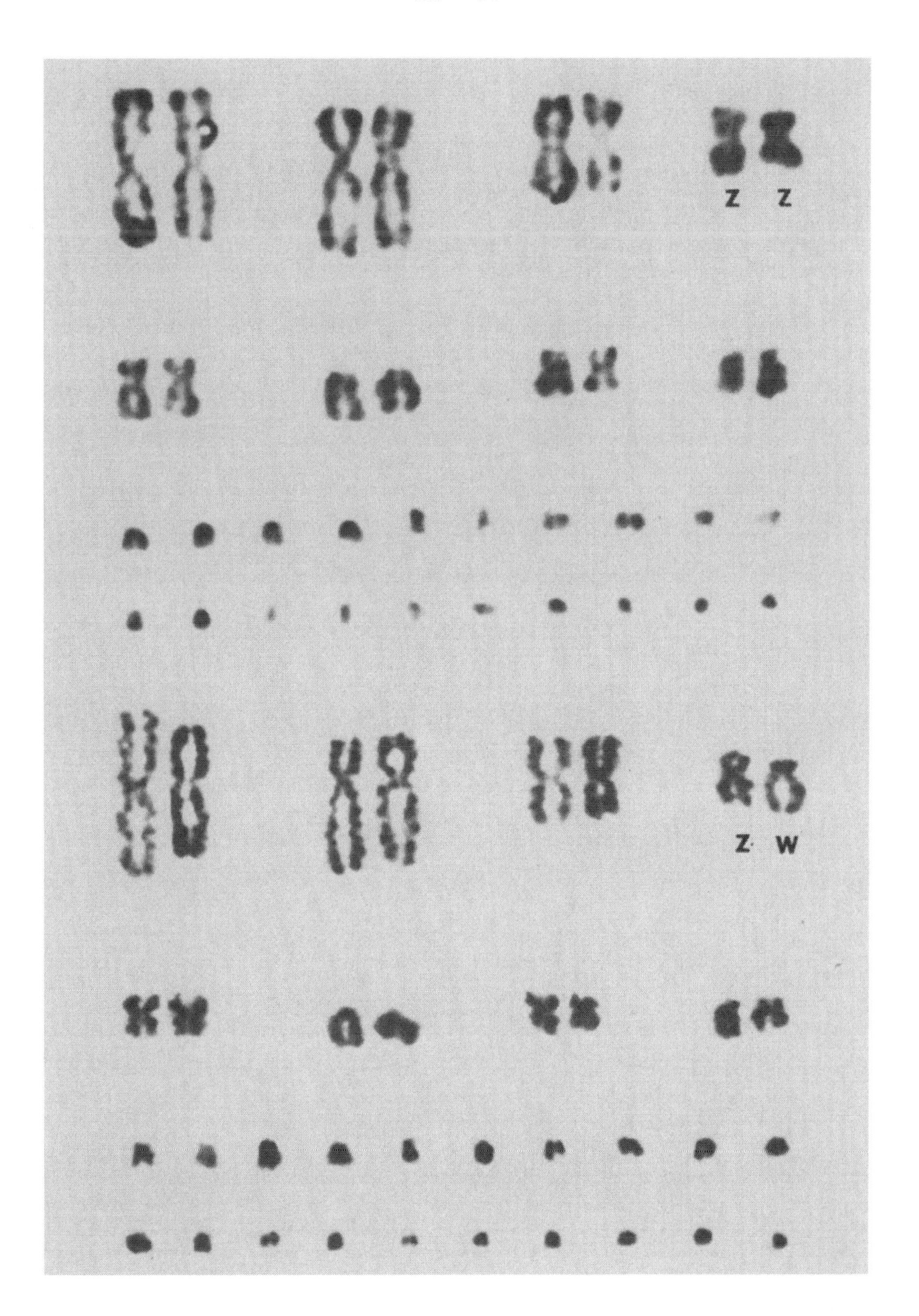

Order: SQUAMATA
Suborder: SERPENTES

Family: COLUBRIDAE

Xenodon neuwiedii (Günther)
(Neuwied's snake; Quiriripitá)
2n = 30

Order: SQUAMATA Family: COLUBRIDAE
Suborder: SERPENTES

Xenodon neuwiedii Günther

(Neuwied's snake; Quiriripitá)

2n = 30

MACROCHROMOSOMES: AUTOSOMES: 12 Metacentrics and submetacentrics
2 Acrocentrics

SEX CHROMOSOMES: Z Metacentric
W Acrocentric

MICROCHROMOSOMES: 14

The karyotypes shown here were obtained from squash preparations of the spleen (female) and intestine (male). The female is heterogametic, the fourth pair of macrochromosomes is heteromorphic in the female (ZW).

The karyotypes are from a female from Espirito Santo and a male from São Paulo, Brazil. The specimens are included in the Collection of the Instituto Butantan.

REFERENCES:

1) Beçak, W. and Beçak, M. L.: Cytotaxonomy and chromosomal evolution in Serpentes. Cytogenetics 8: 247, 1969.

<u>Xenodon</u> <u>neuwiedii</u> Günther

(Neuwied's snake; Quiriripitá)

2n = 30

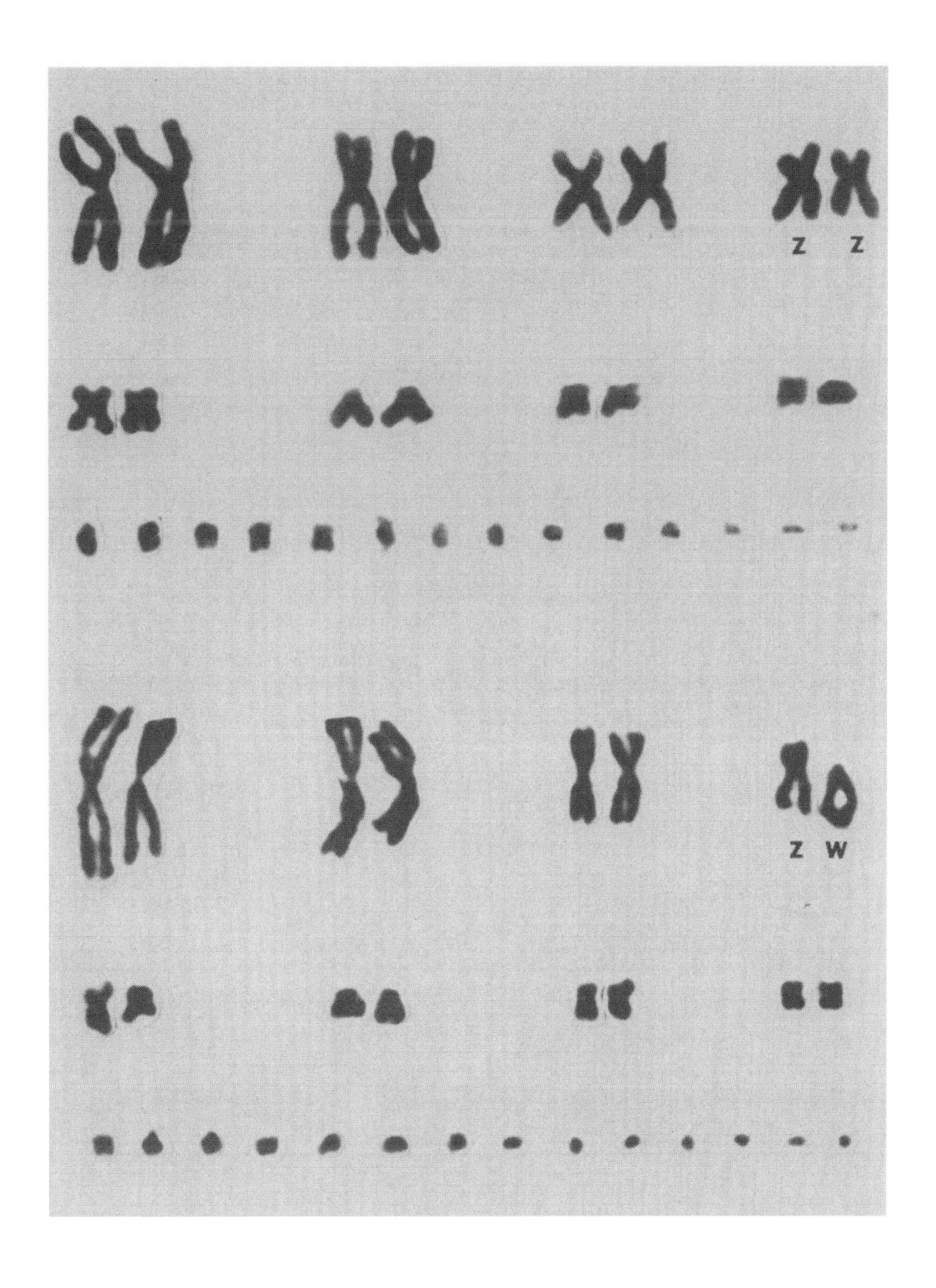

Order: SQUAMATA
Suborder: SAURIA

Family: AMPHISBAENIDAE

Amphisbaena dubia (Müller)

(Worm lizard; Cobra de duas cabeças)

$2n = 26\ (25, 27, 28)$

Order: SQUAMATA Family: AMPHISBAENIDAE
Suborder: SAURIA

Amphisbaena dubia Müller

(Worm lizard; Cobra de duas cabeças)

2n = 26 (2n = 25, 2n = 27, 2n = 28)

MACROCHROMOSOMES: 12 Metacentrics and submetacentrics
2 Acrocentrics

MICROCHROMOSOMES: 12

The karyotypes shown here were obtained from squash preparations of the spleen (female) and intestine (male). Intraindividual polymorphism of number involving macro and microchromosomes was found in the females of the population of São Paulo, Brazil, where these animals were collected.

The karyotypes found are: 2n = 28 (12M + 16m), 2n = 27 (13M + 14m), 2n = 26 (14M + 12m) and 2n = 25 (15M + 10m). The karyotype with 2n = 25 has one macro-acrocentric. Each additional acrocentric coincides with a reduction of two microchromosomes.

The specimens are preserved in the Collection of the Museu de Zoologia da Universidade de São Paulo.

REFERENCES:

1) Beçak, M. L., Beçak, W. and Denaro, L.: Cariologia comparada em oito espécies de lacertílios. Ciência e Cultura supl. 23: 124, 1971.

2) Beçak, M. L., Beçak, W. and Denaro, L.: Chromosome polymorphism, geographical variation and karyotypes of Sauria. Caryologia 25: 313-326, 1972.

Order: SQUAMATA
Suborder: SAURIA

Family: AMPHISBAENIDAE

Amphisbaena dubia Müller

(Worm lizard; Cobra de duas cabeças)

2n = 26 (2n = 25, 2n = 27, 2n = 28)

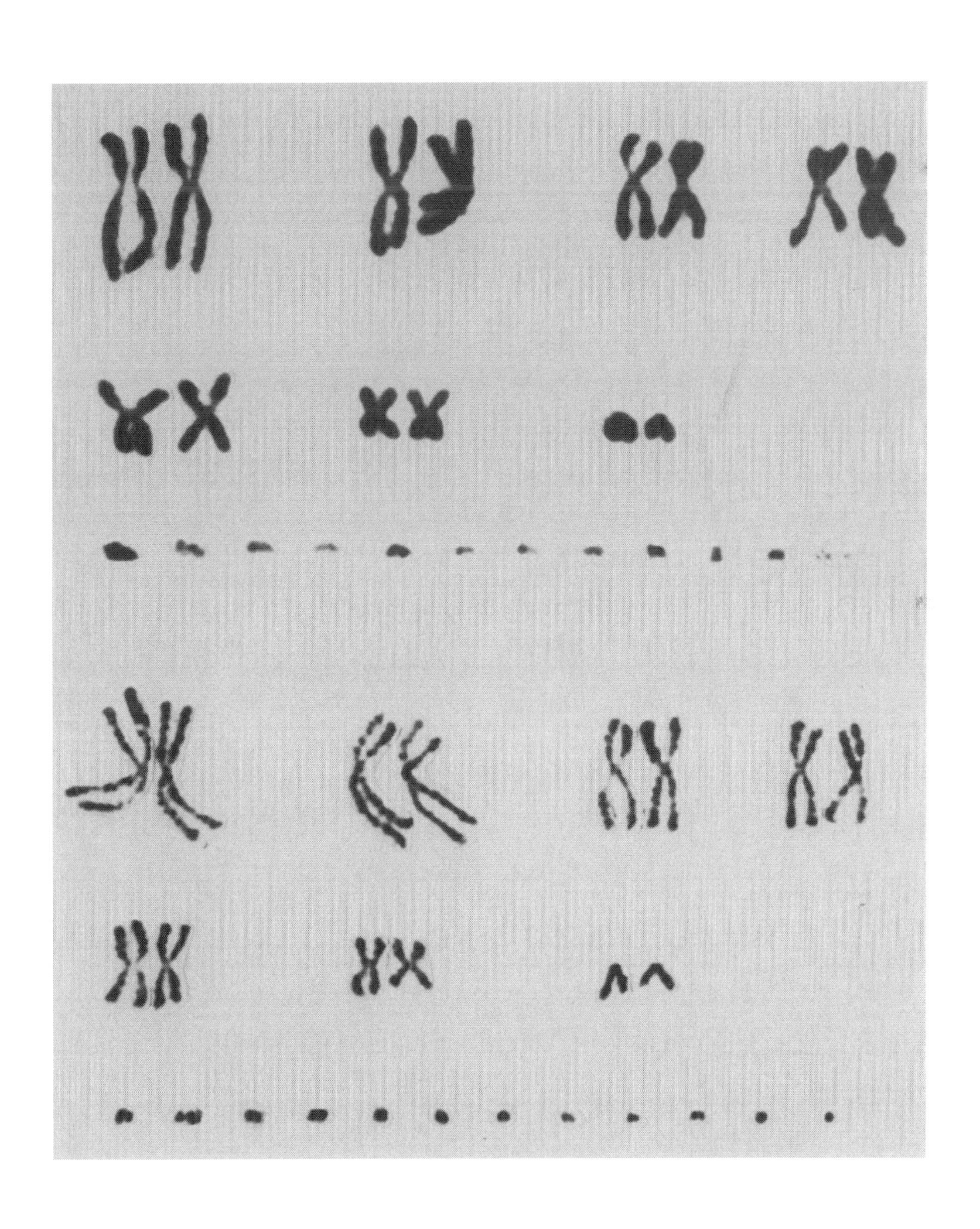

Order: SQUAMATA
Suborder: SAURIA

Family: AMPHISBAENIDAE

Amphisbaena vermicularis (Wagler)

(Worm lizard; Cobra de duas cabeças)

2n = 44

Order: SQUAMATA
Suborder: SAURIA

Family: AMPHISBAENIDAE

Amphisbaena vermicularis Wagler

(Worm lizard; Cobra de duas cabeças)

2n = 44

MACROCHROMOSOMES: 2 Metacentrics
20 Acrocentrics

MICROCHROMOSOMES: 22

The karyotypes were obtained from squash preparations of the intestine. Both male and female have similar karyotypes; sex chromosome hetemorphism was not detected in either sex.

The animals were collected in Maranhão Brazil. The specimens are included in the Collection of the Museu de Zoologia da Universidade de São Paulo.

REFERENCES:

1) Beçak, M. L., Beçak, W. and Napoleone, L. F.: Personal communication.

Order: SQUAMATA
Suborder: SAURIA

Family: AMPHISBAENIDAE

<u>Amphisbaena</u> <u>vermicularis</u> Wagler

(Worm lizard; Cobra de duas cabeças)

2n = 44

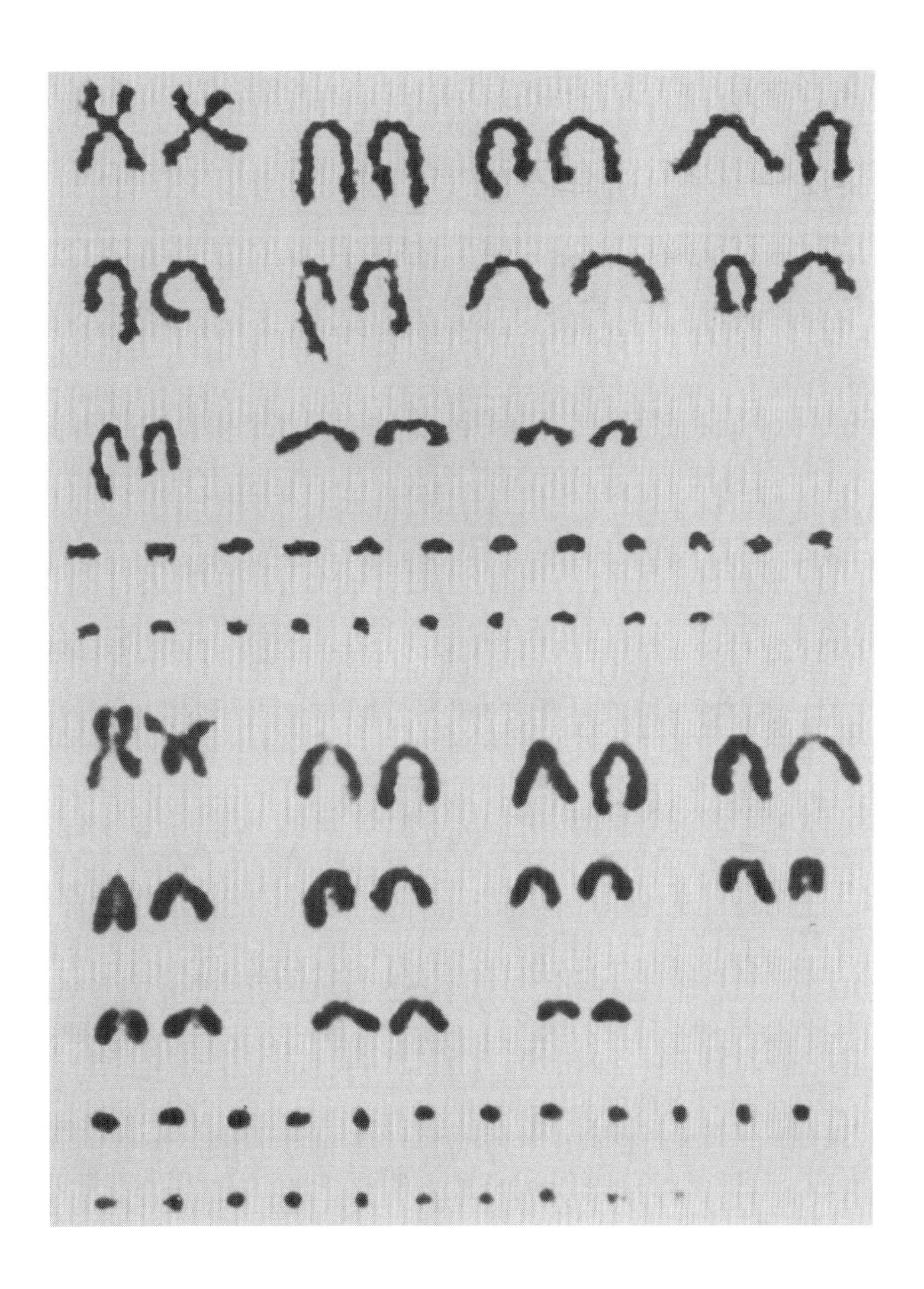

Order: SQUAMATA
Suborder: SAURIA

Family: AMPHISBAENIDAE

Leposternon microcephalum (Wagler)

(Cobra de duas cabeças)

2n = 34

Order: SQUAMATA
Suborder: SAURIA

Family: AMPHISBAENIDAE

Leposternon microcephalum Wagler

(Cobra de duas cabeças)

2n = 34

MACROCHROMOSOMES: 12 Metacentrics and submetacentrics

MICROCHROMOSOMES: 22

The karyotypes presented here were obtained from squash preparation of the intestine (female) and tissue culture (male). Sex chromosome heteromorphism was not detected in either sex.

The male karyotype was kindly provided by Dr. Carl Gans of the University of Michigan, U.S.A. in which collection it is included. The female specimen captured in São Paulo, Brazil, is preserved in the Collection of the Museu de Zoologia da Universidade de São Paulo.

REFERENCES:

1) Huang, C. C., Clark, H. F. and Gans, C.: Karyological studies on fifteen forms of Amphisbaenians (Amphisbaenia, Reptilia). Chromosoma 22: 1, 1967.

2) Beçak, M. L., Beçak, W. and Denaro, L.: Chromosome polymorphism, geographical variation and karyotypes of Sauria. Caryologia 25: 313-326, 1972.

Order: SQUAMATA
Suborder: SAURIA

Family: AMPHISBAENIDAE

Leposternon microcephalum Wagler

(Cobra de duas cabeças)

2n = 34

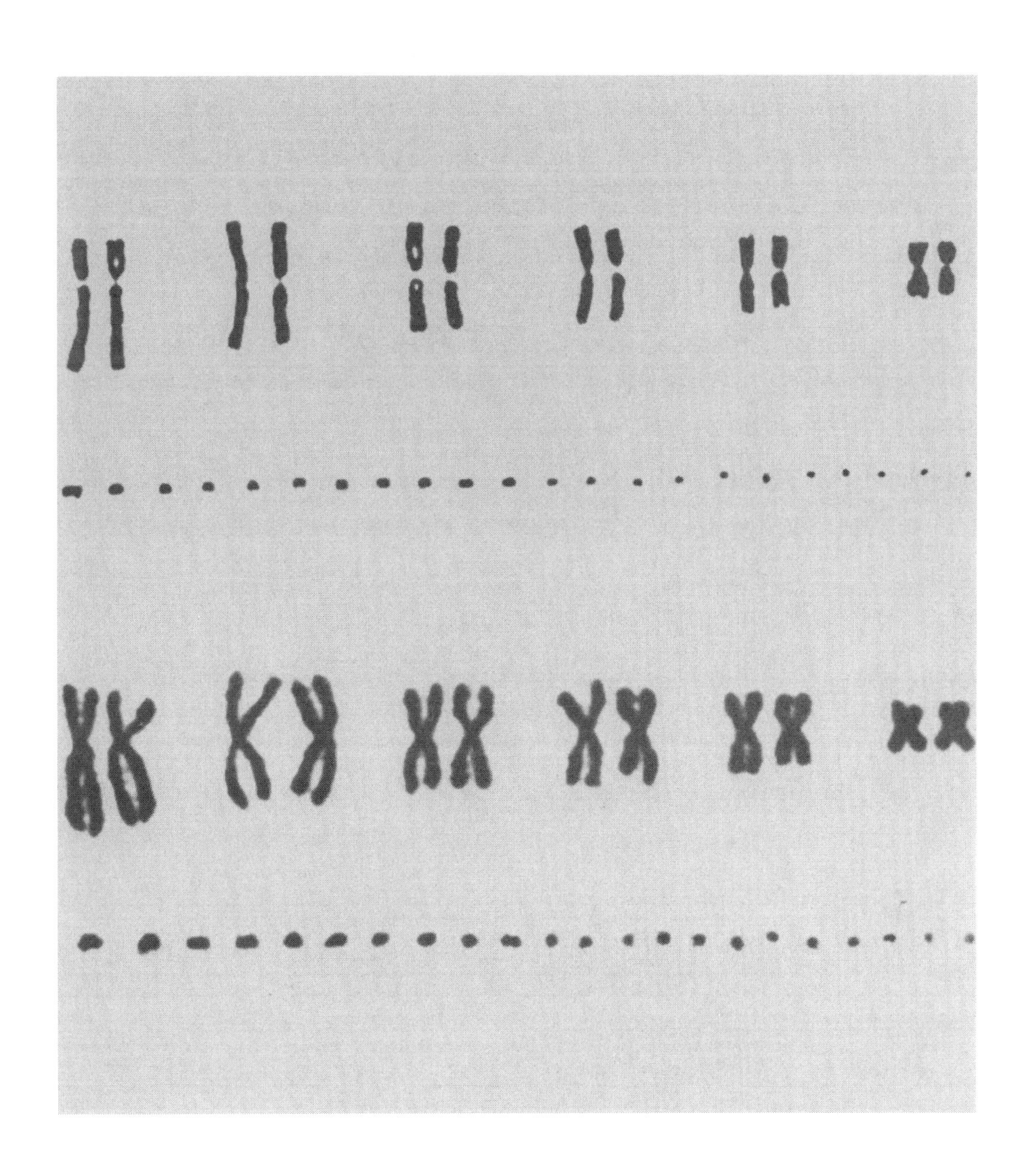

Order: SQUAMATA
Suborder: SAURIA

Family: SCINCIDAE

Mabuya mabouya mabouya (Lacépède)

(Mabuya; Lagartixa)

2n = 30

Order: SQUAMATA Family: SCINCIDAE
Suborder: SAURIA

Mabuya mabouya mabouya (Lacépède)

(Mabuya; Lagartixa)

2n = 30

MACROCHROMOSOMES: 16 Metacentrics and submetacentrics

MICROCHROMOSOMES: 14

The karyotypes presented here were obtained from squash preparations of the intestine (female) and the gonads (male). Sex chromosome heteromorphism was not detected in either sex.

The male specimen was captured in Maranhão and the female specimen in São Paulo, Brazil. Both specimens are preserved in the Collection of the Museu de Zoologia da Universidade de São Paulo.

REFERENCES:

1) Beçak, M. L., Beçak, W. and Denaro, L.: Cariologia comparada em oito espécies de lacertílios. Ciência e Cultura supl. 23: 124, 1971.

2) Beçak, M. L., Beçak, W. and Denaro, L.: Chromosome polymorphism, geographical variation and karyotypes of Sauria. Caryologia 25: 313-326, 1972.

Order: SQUAMATA
Suborder: SAURIA

Family: SCINCIDAE

<u>Mabuya mabouya mabouya</u> (Lacépède)

(Mabuya; Lagartixa)

2n = 30

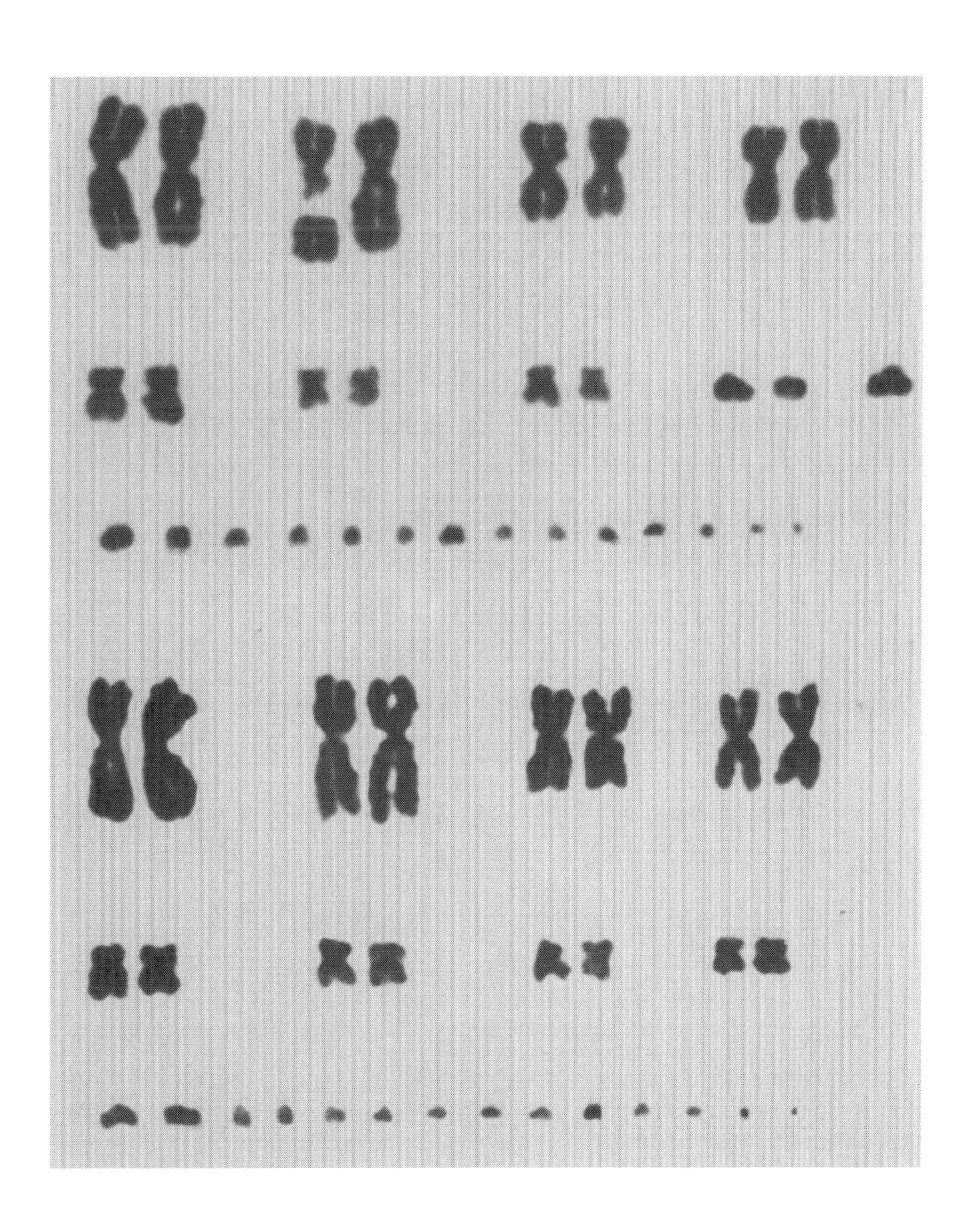

Order: SQUAMATA
Suborder: SAURIA

Family: ANGUIDAE

Ophiodes striatus (Spix)

(Glass snake; Cobra de vidro)

2n = 36

Order: SQUAMATA
Suborder: SAURIA

Family: ANGUIDAE

Ophiodes striatus (Spix)

(Glass snake; Cobra de vidro)

2n = 36

MACROCHROMOSOMES: 12 Metacentrics and submetacentrics

MICROCHROMOSOMES: 24

The karyotypes shown here were obtained from squash preparation of the spleen (male) and short term blood culture (female). Both male and female have similar karyotypes; sex chromosome heteromorphism was not detected in either sex.

The animals were collected in Brazil. The specimens are preserved in the Collection of the Museu de Zoologia da Universidade de São Paulo.

REFERENCES:

1) Beçak, M. L., Beçak, W. and Denaro, L.: Cariologia comparada em oito espécies de lacertílios. Ciência e Cultura supl. 23: 124, 1971.

2) Beçak, M. L., Beçak, W. and Denaro, L.: Chromosome polymorphism, geographical variation and karyotypes of Sauria. Caryologia 25: 313-326, 1972.

Order: SQUAMATA
Suborder: SAURIA

Family: ANGUIDAE

<u>Ophiodes striatus</u> (Spix)

(Glass snake; Cobra de vidro)

2n = 36

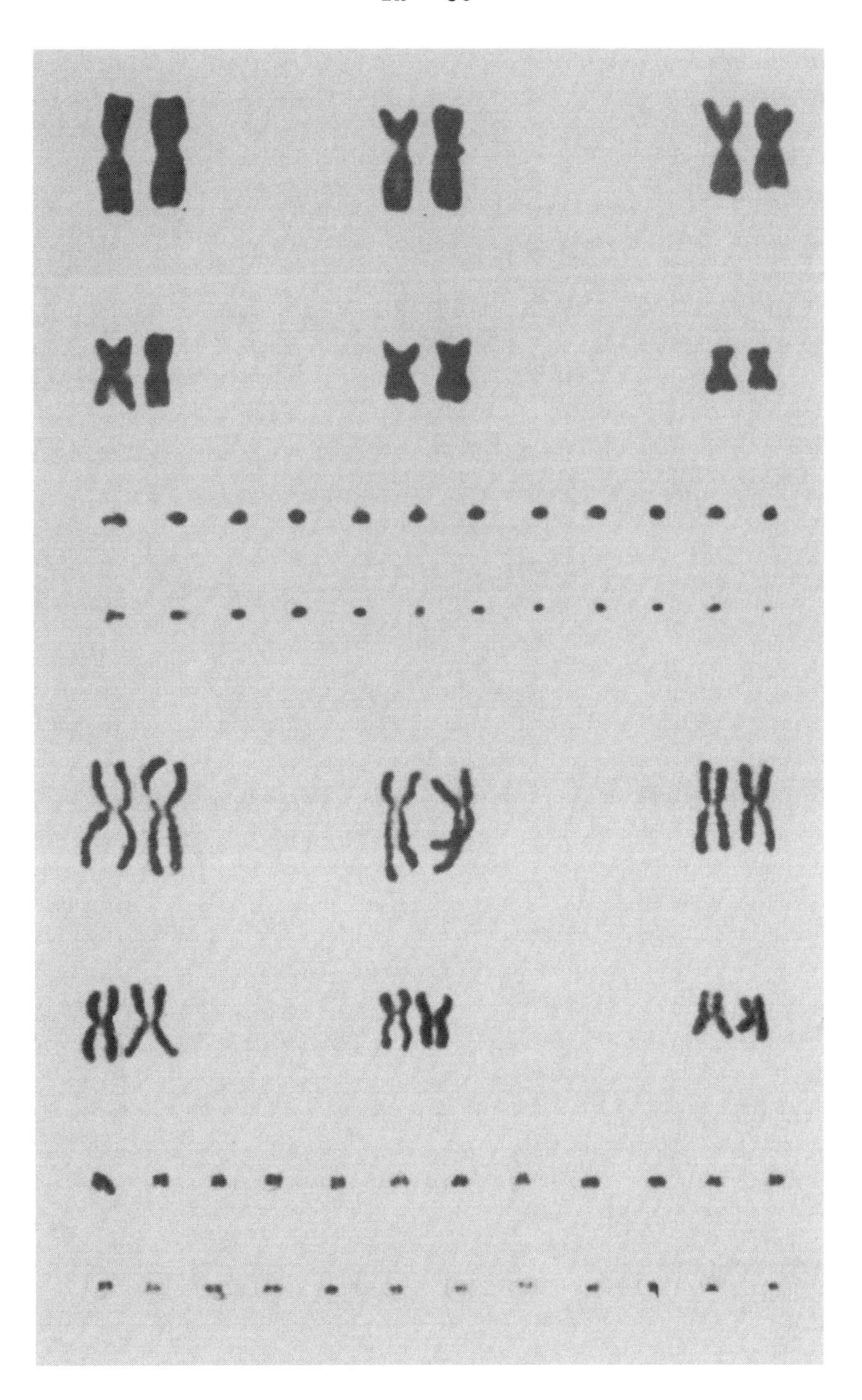

Order: SQUAMATA
Suborder: SAURIA

Family: IGUANIDAE

Tropidurus torquatus (Wied)

(Calango)

$2n = 36$ (37, 38)

Order: SQUAMATA Family: IGUANIDAE
Suborder: SAURIA

Tropidurus torquatus (Wied)

(Calango)

2n = 36 (2n = 37, 2n = 38)

MACROCHROMOSOMES: 12 Metacentrics and submetacentrics

MICROCHROMOSOMES: 24

The karyotypes were obtained by squash preparations of intestine. An intraindividual and intrapopulational karyotype variation in number apparently due to supernumerary chromosomes was observed in the iguanids from Minas Gerais and Maranhão, Brazil: 2n = 36 (12 + 24m), 2n = 37 (13M + 24m) and 2n = 38 (14M + 24m). The additional macrochromosome is submetacentric and the smallest of the karyotype.

An heteromorphic autosomal pair was found in another population of iguanids, from São Paulo, Brazil. The dimorphic pair consists of a large submetacentric and a smaller metacentric.

The animals used for the karyotypes shown here were collected in Maranhão, Brazil. The specimens are included in the Collection of the Museu de Zoologia da Universidade de São Paulo.

REFERENCES:

1) Beçak, M. L., Beçak, W. and Denaro, L.: Cariologia comparada em oito espécies de lacertílios. Ciência e Cultura supl. 23: 124, 1971.

2) Beçak, M. L., Beçak, W. and Denaro, L.: Chromosome polymorphism, geographical variation and karyotypes of Sauria. Caryologia 25: 313-326, 1972.

3) Gorman, G. C., Atkins, L. and Holzinger, T.: New karyotypic data on 15 genera of lizards in the family Iguanidae, with discussion of taxonomic and cytological implications. Cytogenetics 6: 286, 1967.

Order: SQUAMATA
Suborder: SAURIA

Family: IGUANIDAE

<u>Tropidurus torquatus</u> (Wied)

(Calango)

2n = 36 (2n = 37, 2n = 38)

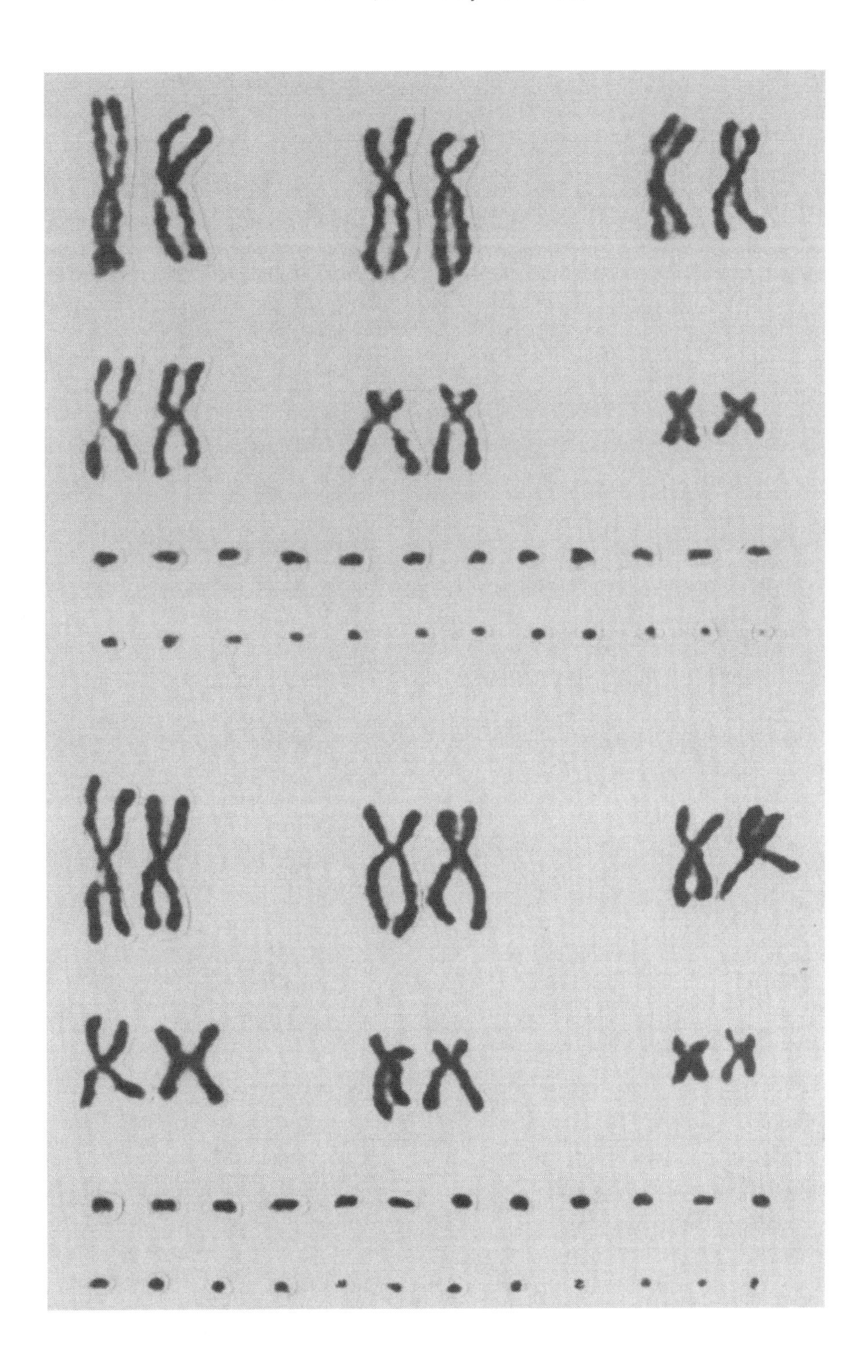

Order: RHEIFORMES

Family: RHEIDAE

Rhea americana

(Rhea)

2n = 82

Order: RHEIFORMES Family: RHEIDAE

Rhea americana (Rhea)

2n = 82

AUTOSOMES: 12 or more submetacentrics, metacentrics or subtelocentrics
The remainder are presumed to be acrocentrics

SEX CHROMOSOMES: Z Acrocentric
W Acrocentric or subtelocentric

The karyotypes were prepared from feather pulp of specimens obtained from the Como Zoo, St. Paul, Minnesota. Takagi et al. describe a slightly hetromorphic acrocentric pair (Z and W) which they assign to size six position.

REFERENCES:

1) Takagi, N., M. Itoh and M. Sasaki. 1972. Chromosome studies in four species of Ratitae (Aves). Chromosoma (Berl.) 36, 281-291.

<u>Rhea americana</u> (Rhea)

2n = 82

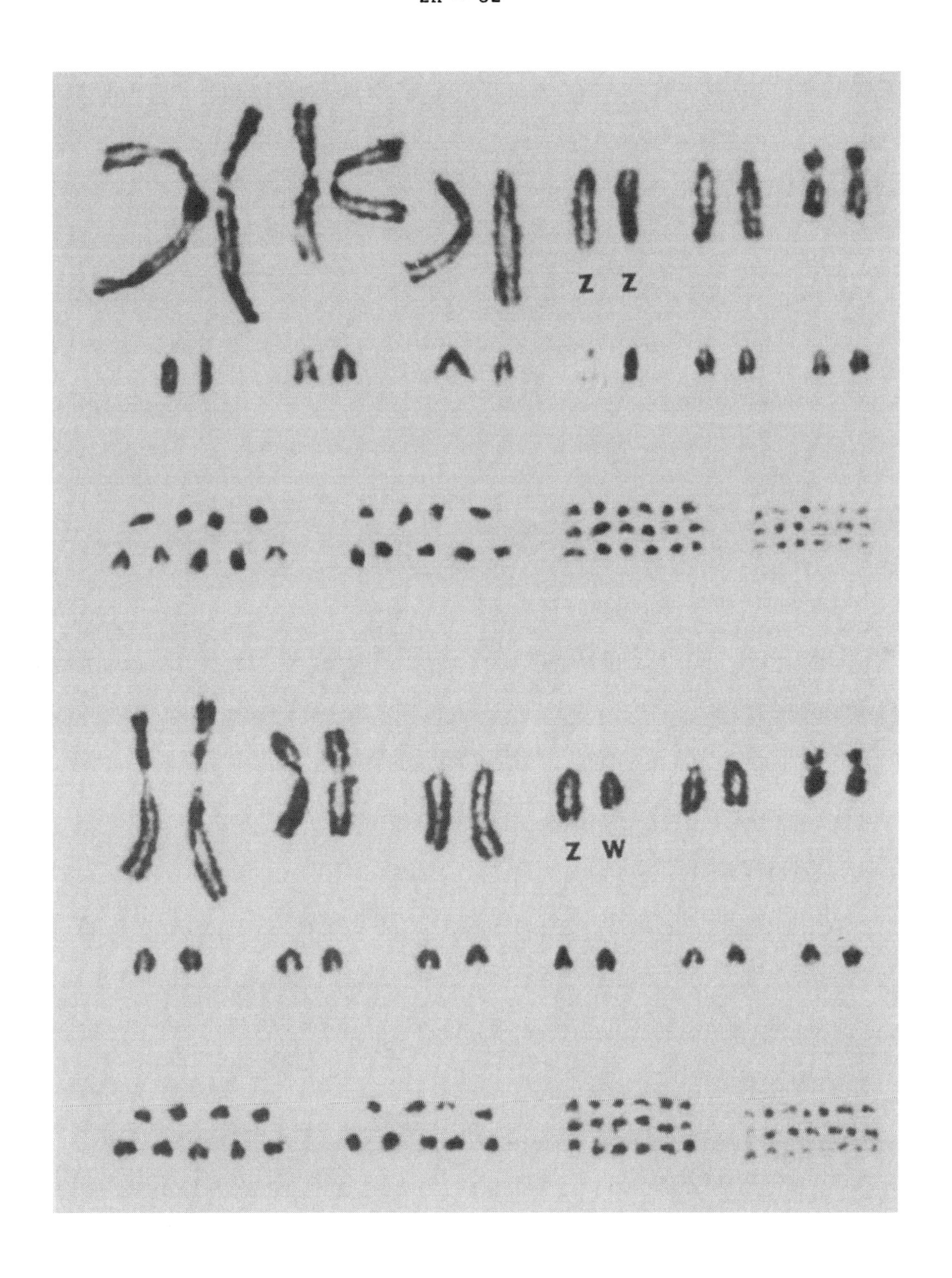

Order: CHARADRIFORMES

Family: CHARADRIDAE

Charadrius vociferous

(Killdeer)

$2n = 76$

Order: CHARADRIIFORMES | Family: CHARADRIIDAE

Charadrius vociferus (Killdeer)

2n = 76

AUTOSOMES: 10 Submetacentrics, metacentrics, or subtelocentrics
The remainder are presumed to be acrocentric

SEX CHROMOSOMES: Z Metacentric
W Acrocentric or subtelocentric

The mitotic material was obtained from feather pulp of specimens trapped in the vicinity of St. Paul, Minnesota.

Charadrius vociferus (Killdeer)

2n = 76

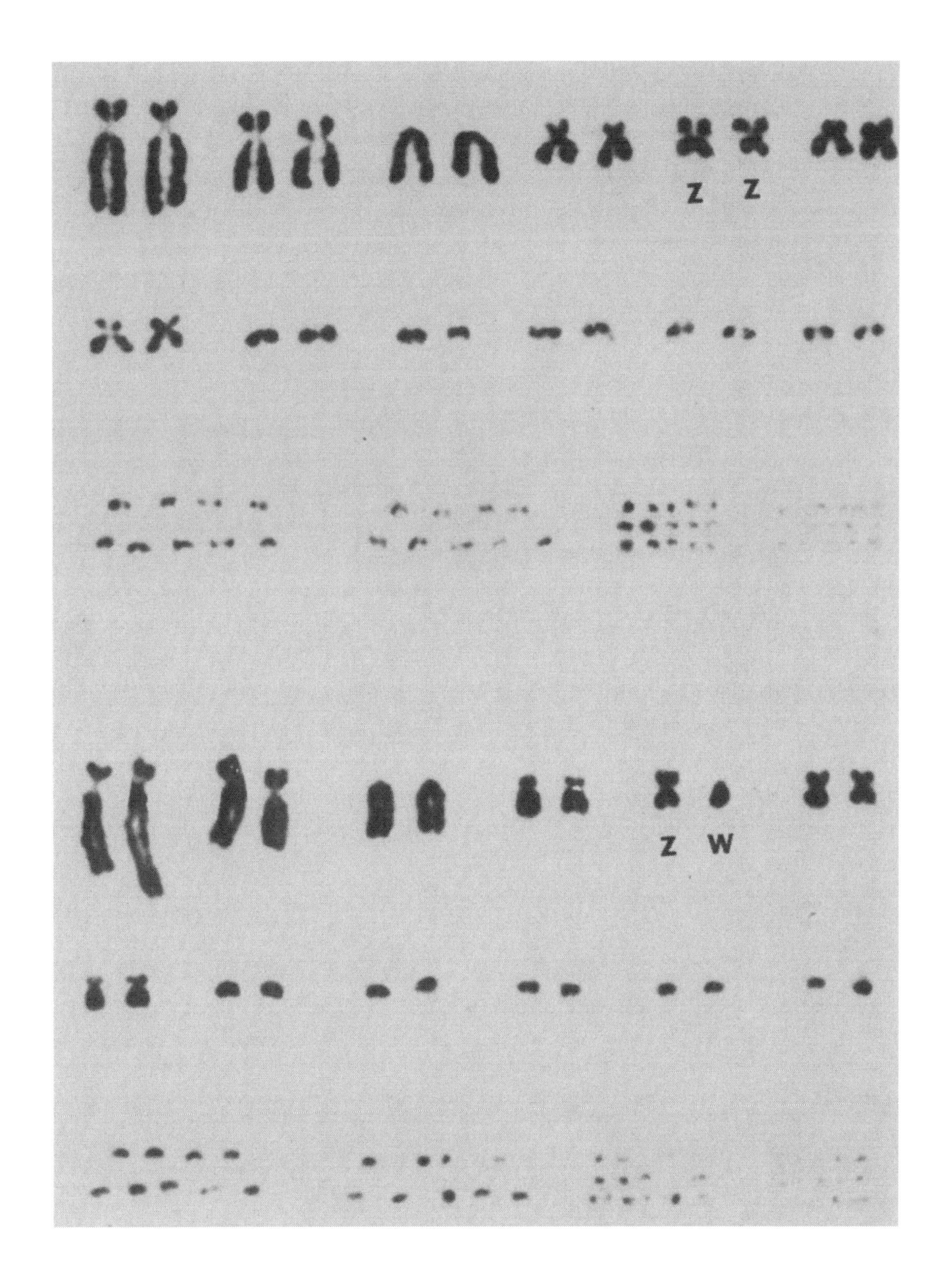

Order: PASSERIFORMES

Family: CORVIDAE

Corvus corax

(Raven)

$2n = 78$

Order: PASSERIFORMES   Family: CORVIDAE

Corvus corax (Raven)

2n = 78

AUTOSOMES: 14 or more submetacentrics, metacentrics or subtelocentrics

SEX CHROMOSOMES: Z Acrocentric or subtelocentric
W Subtelocentric or submetacentric

The karyotypes shown were prepared from feather pulp. Specimens were obtained from the Como Zoo, St. Paul, Minnesota.

Corvus corax (Raven)

2n = 78

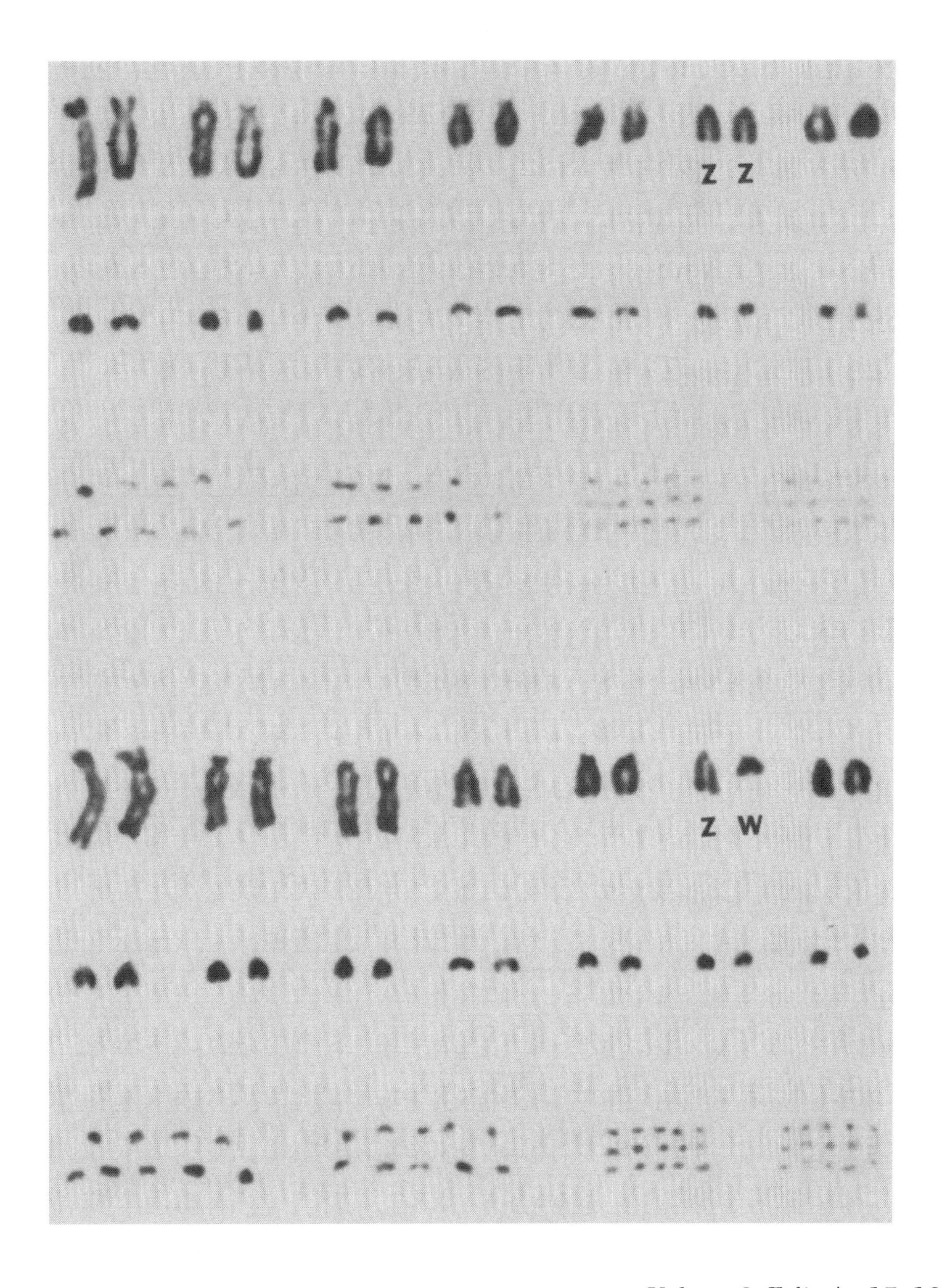

Order: STRIGIFORMES

Family: STRIGIDAE

Bubo virginianus

(Great horned owl)

2n = 82

Order: STRIGIFORMES Family: STRIGIDAE

Bubo virginianus (Great horned owl)

2n = 82

AUTOSOMES: 10 or more submetacentrics, metacentrics or subtelocentrics
The remainder are presumed to be acrocentrics

SEX CHROMOSOMES: Z Metacentric
W Metacentric or submetacentric

The karyotypes were prepared from feather pulp obtained from specimens in the St. Paul, Minnesota vicinity.

REFERENCES:

1) Krishan, Atwar, G. J. Haiden, and R. N. Shoffner. 1965. Mitotic chromosomes and the W-sex chromosome of the Great Horned Owl (*Bubo v. virginianus*). Chromosoma (Berl.) 17, 258-263.

Bubo virginianus (Great horned owl)

2n = 82

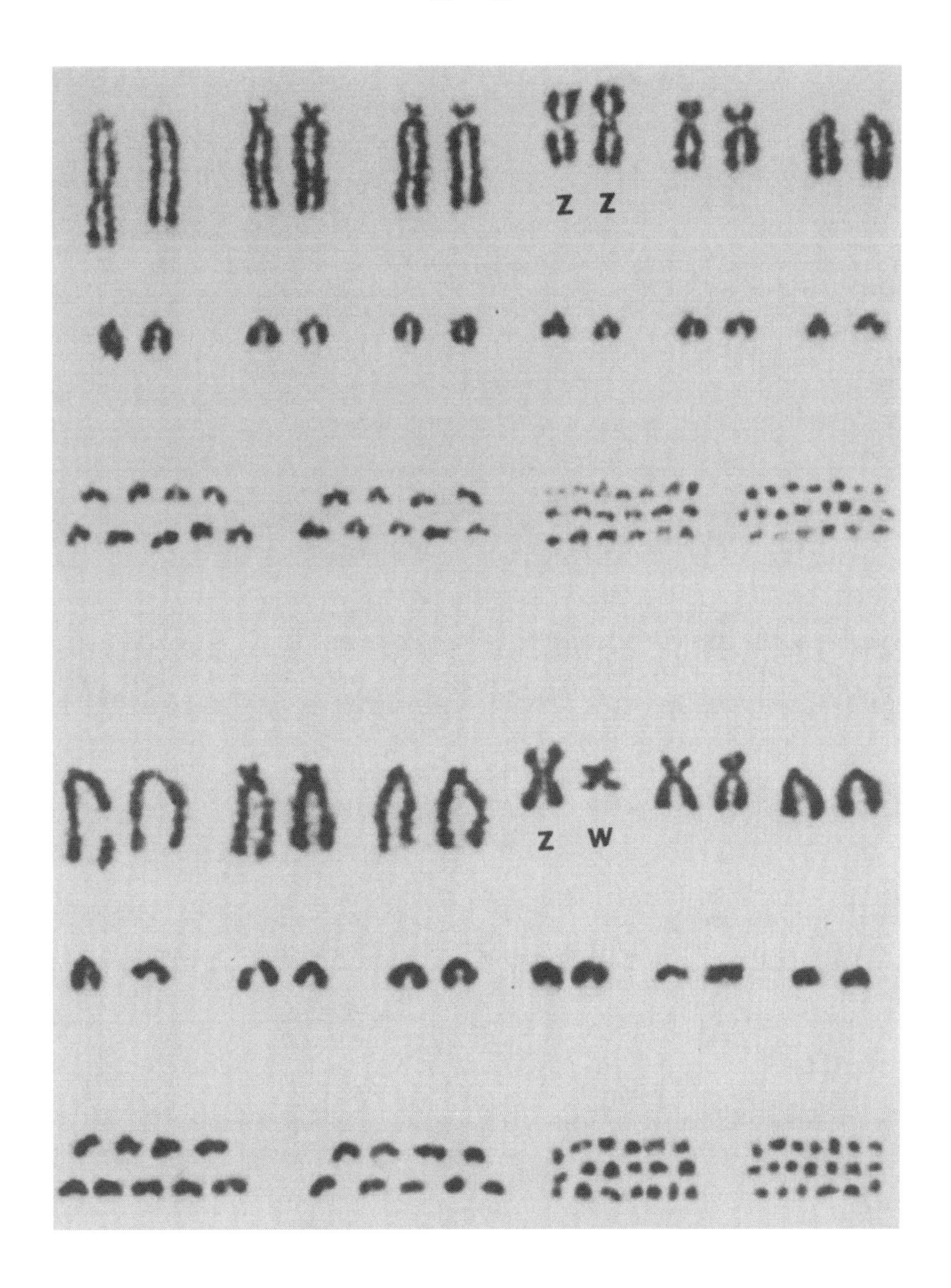

Order: APODIFORMES

Family: TROCHILIDAE

Calypte anna

(Anna's humming bird)

2n = 74

Order: APODIFORMES Family: TROCHILIDAE

Calypte anna (Anna's humming bird)

2n = 74

AUTOSOMES: 24 or more submetacentrics, subtelocentrics or metacentrics
The remainder are presumed to be acrocentrics

SEX CHROMOSOMES: Z Submetacentric
W Subtelocentric or acrocentric

The idiogram negatives for this species were very graciously supplied by Ms. Ingrid Benirschke, La Jolla, California. The mitotic figures were obtained from cultures of lung and kidney tissue.

REFERENCES:

1) Benirschke, Ingrid. 1971. The chromosomes of Anna's hummingbird. Unpublished data.

Calypte anna (Anna's humming bird)

2n = 74

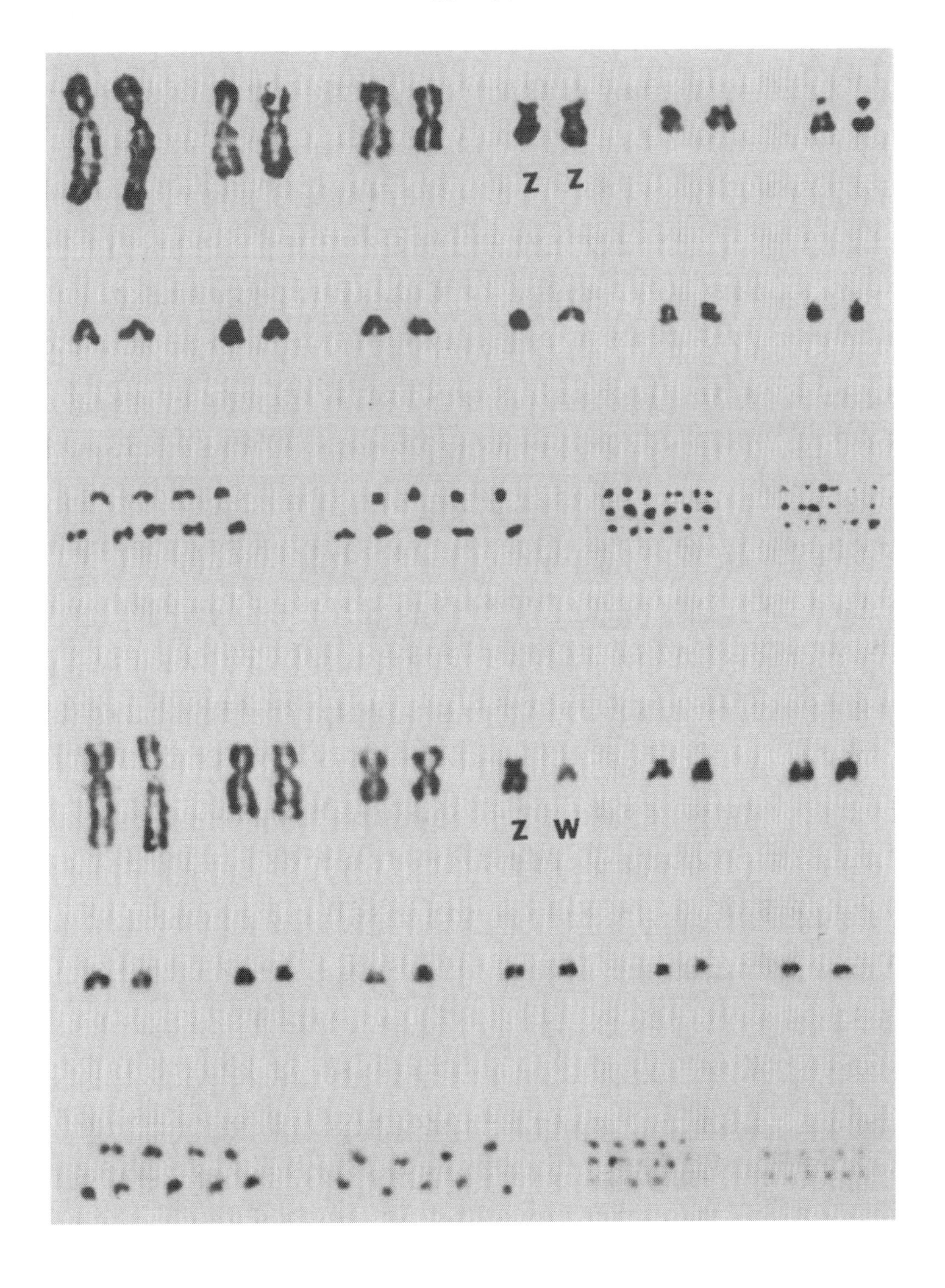

Order: GALLIFORMES

Family: PHASIANIDAE

Lophortyx gambelli

(Gambell's quail)

2n = 80

Order: GALLIFORMES Family: PHASIANIDAE

Lophortyx gambelli (Gambell's quail)

2n = 80

AUTOSOMES: At least 16 metacentrics, submetacentrics or subtelocentrics
The remainder are presumed to be acrocentrics

SEX CHROMOSOMES: Z Submetacentric
W Subtelocentric or acrocentric

Specimens were kindly supplied to us by Dr. Paul Johnsgard of the University of Nebraska and by a local breeder in the St. Paul, Minnesota area. Mitotic material was obtained from feather pulp.

Lophortyx gambelli (Gambell's quail)

2n = 80

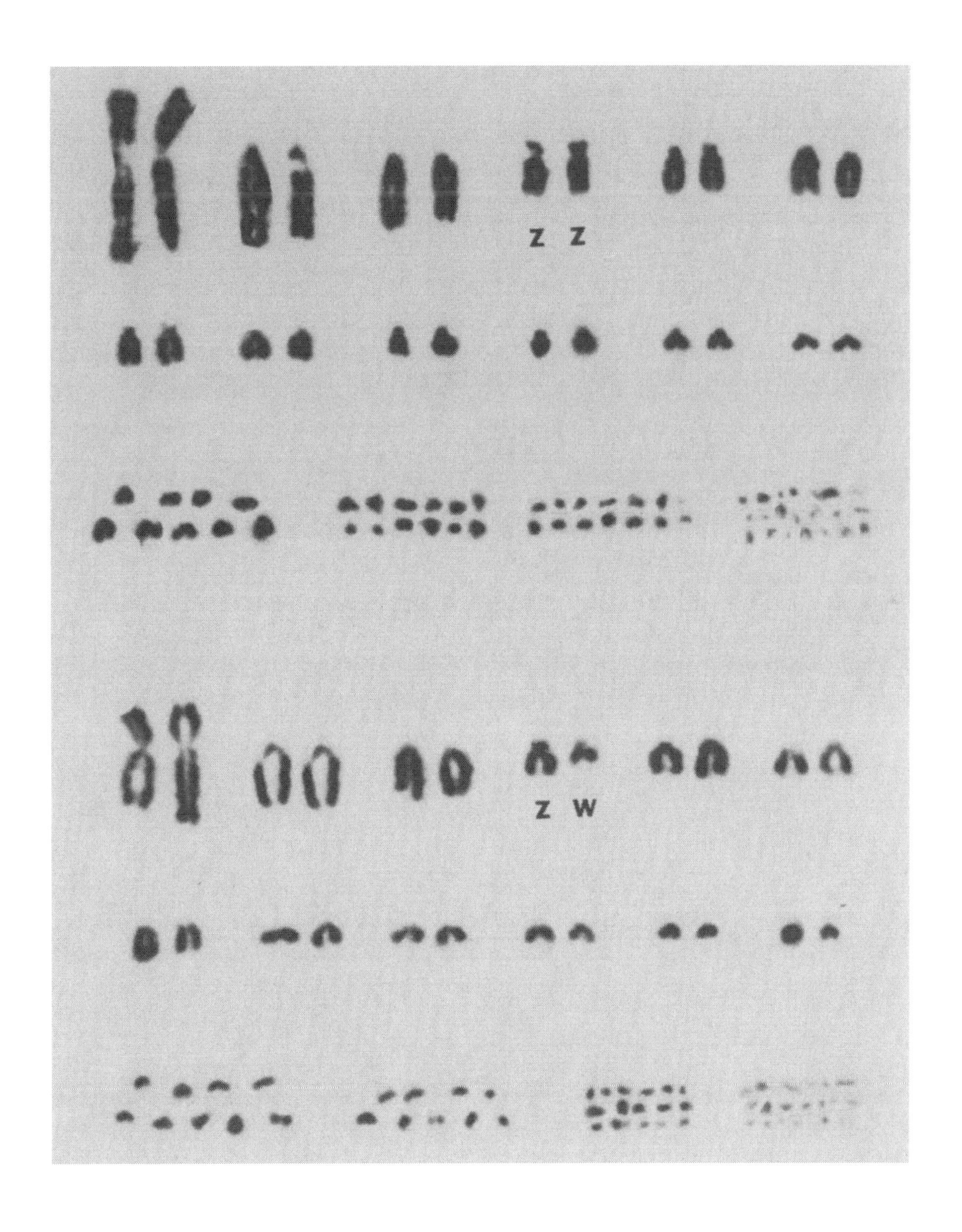

Order: ANSERIFORMES

Family: ANATIDAE
Tribe: CAIRININI

Aix galericulata

(Mandarin duck)

2n = 90

Order: ANSERIFORMES

Family: ANATIDAE
Tribe: CAIRININI

Aix galericulata (Mandarin duck)

2n = 90

AUTOSOMES: All acrocentric

SEX CHROMOSOMES: Z Acrocentric or subtelocentric
W Acrocentric

Several specimens were supplied to us by Mr. Forrest Lee of the Northern Prairie Wildlife Research Station, U.S.F.W.S., Jamestown, N.D. Mitotic plates were obtained from feather pulp and leucocyte cultures. The unusual morphology and larger number of chromosomes of the Mandarin is considerably different from the Wood Duck (*Aix sponsa*) (Av-22).

Aix galericulata (Mandarin duck)

2n = 90

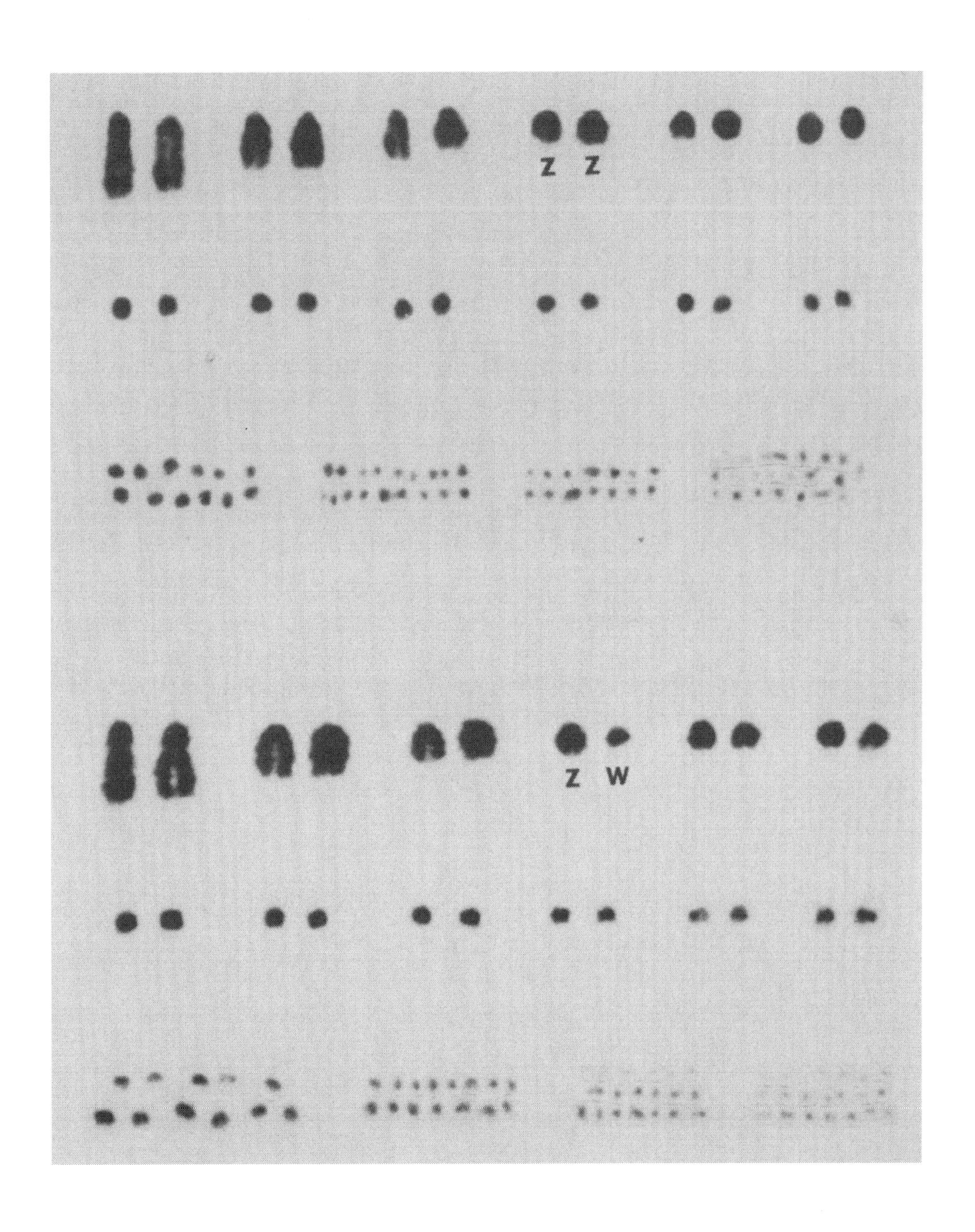

Order: ANSERIFORMES

Family: ANATIDAE
Tribe: CAIRININI

Aix sponsa

(Wood duck)

2n = 80

Order: ANSERIFORMES

Family: ANATIDAE
Tribe: CAIRININI

Aix sponsa (Wood duck)

2n = 80

AUTOSOMES: 4 Submetacentrics
The remainder are presumed to be acrocentrics

SEX CHROMOSOMES: Z Acrocentric
W Acrocentric

Specimens were obtained from local breeders in the St. Paul, Minnesota area and from the Northern Prairie Wildlife Research Station, U.S.F.W.S., Jamestown, H.D. Feather pulp was the primary source of mitotic plates.

Aix sponsa (Wood duck)

2n = 80

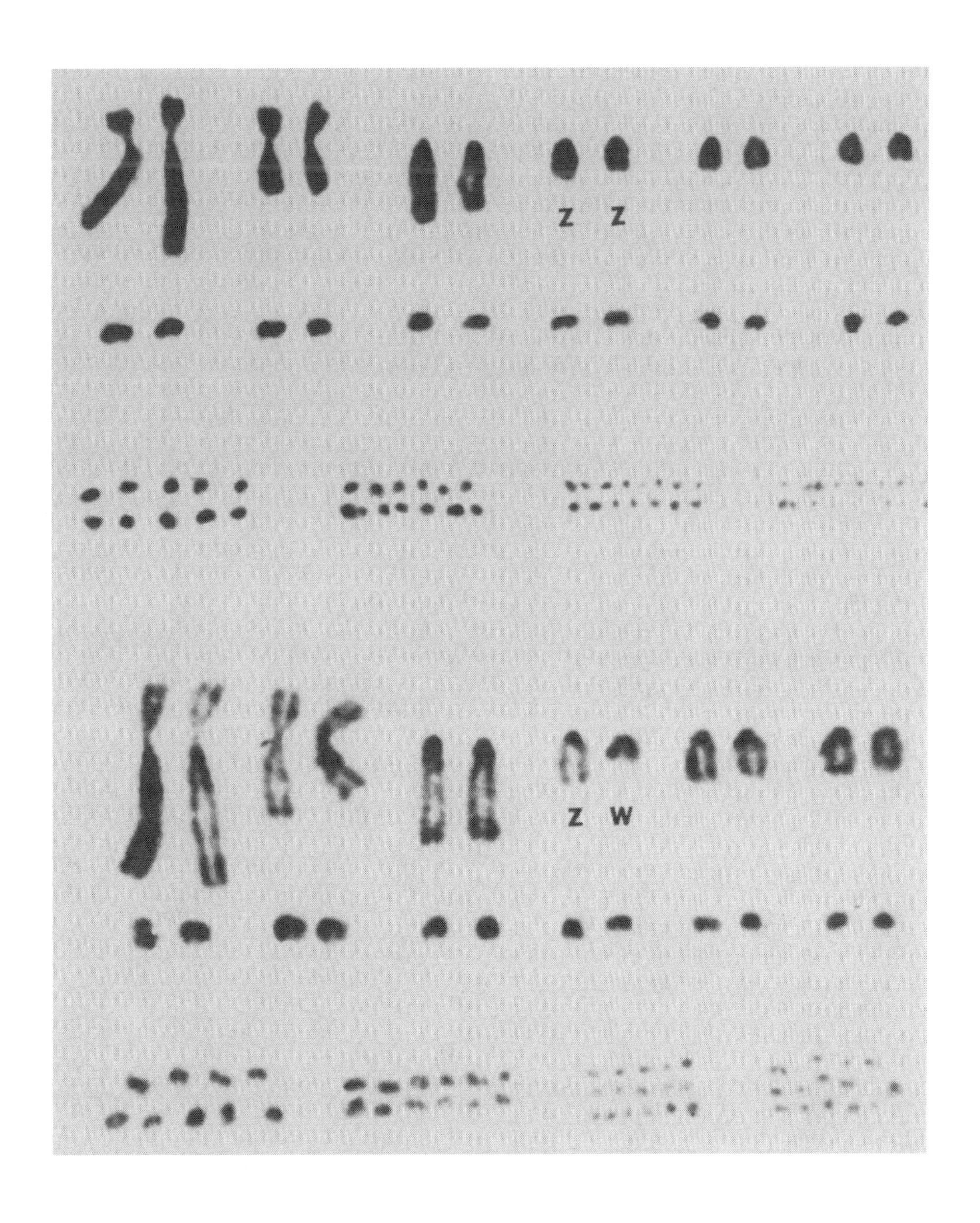

Order: ANSERIFORMES

Family: ANATIDAE

Tribe: CAIRININI

Cairina moschata

(Muscovy duck)

$2n = 78$

Order: ANSERIFORMES

Family: ANATIDAE
Tribe: CAIRININI

<u>Cairina</u> <u>moschata</u> (Muscovy duck)

2n = 78

AUTOSOMES: 4 Submetacentrics
The remainder are presumed to be acrocentrics

SEX CHROMOSOMES: Z Acrocentric
W Acrocentric

Several specimens of domestic varieties were obtained from breeders in the vicinity of St. Paul, Minnesota. Both feather pulp and leucocyte cultures were used as a source of mitotic plates.

REFERENCES:

1) Mott, C. L., L. H. Lockhart, and R. H. Rigdon. 1968. Chromosomes of the sterile hybrid duck. Cytogenetics 7, 403-412.

Order: ANSERIFORMES

Family: ANATIDAE
Tribe: CAIRININI

Cairina moschata (Muscovy duck)

2n = 78

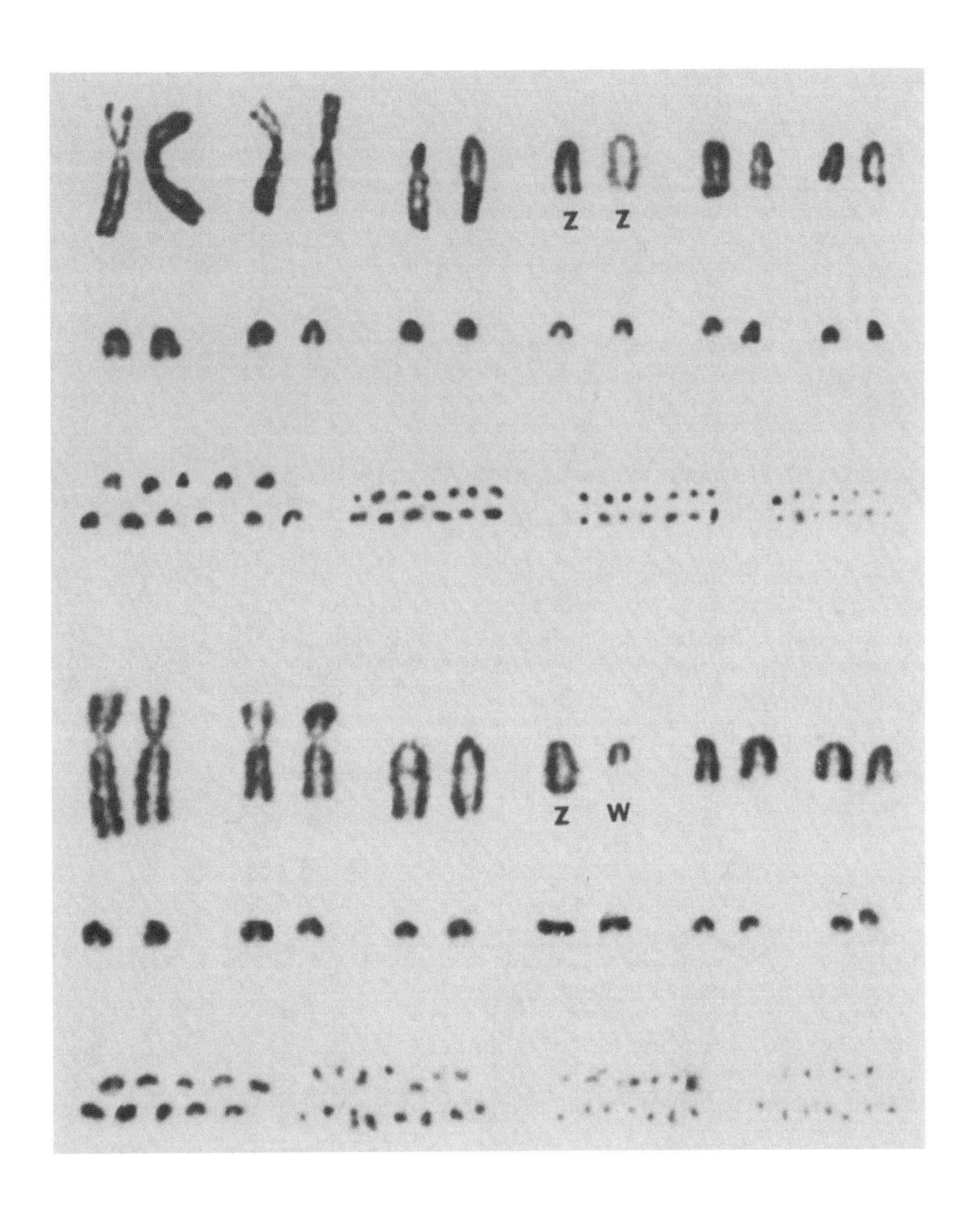

Order: ANSERIFORMES

Family: ANATIDAE
Tribe: DENDROCYGNINI

Dendrocygna bicolor

(Fulvous tree duck)

2n = 80

Order: ANSERIFORMES

Family: ANATIDAE
Tribe: DENDROCYGNINI

Dendrocygna bicolor (Fulvous tree duck)

2n = 80

AUTOSOMES: 4 Submetacentrics
The remainder are presumed to be acrocentric

SEX CHROMOSOMES: Z Acrocentric
W Acrocentric

Single pair specimens were supplied by Mr. Forrest Lee of the Northern Prairie Wildlife Research Station, U.S.F.W.S., Jamestown, N.D. Feather pulp was used as the source of mitotic plates.

Order: ANSERIFORMES

Family: ANATIDAE
Tribe: DENDROCYGNINI

Dendrocygna bicolor (Fulvous tree duck)

2n = 80

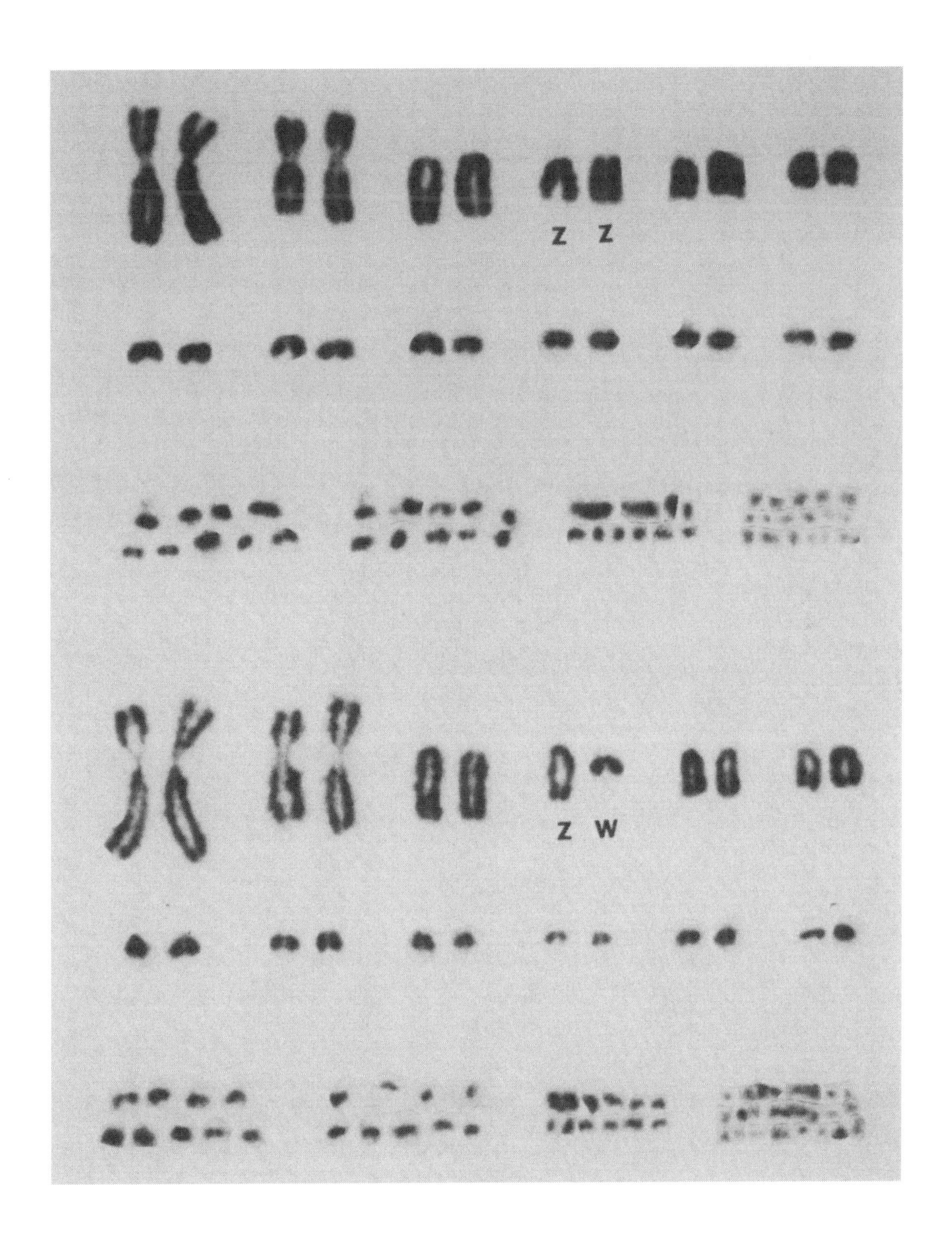

Order: ANSERIFORMES

Family: ANATIDAE
Tribe: ANATINI

Anas acuta

(Pintail duck)

$2n = 82$

Order: ANSERIFORMES

Family: ANATIDAE
Tribe: ANATINI

Anas acuta (Pintail duck)

2n = 82

AUTOSOMES: 4-6 Submetacentrics
The remainder are presumed to be acrocentrics

SEX CHROMOSOMES: Z Subtelocentric
W Acrocentric

Single pair specimens were supplied by Mr. Forrest Lee of the Northern Prairie Wildlife Research Station, U.S.F.W.S., Jamestown, N.D. Mitotic plates were obtained from feather pulp.

Order: ANSERIFORMES

Family: ANATIDAE
Tribe: ANATINI

Anas acuta (Pintail duck)

2n = 82

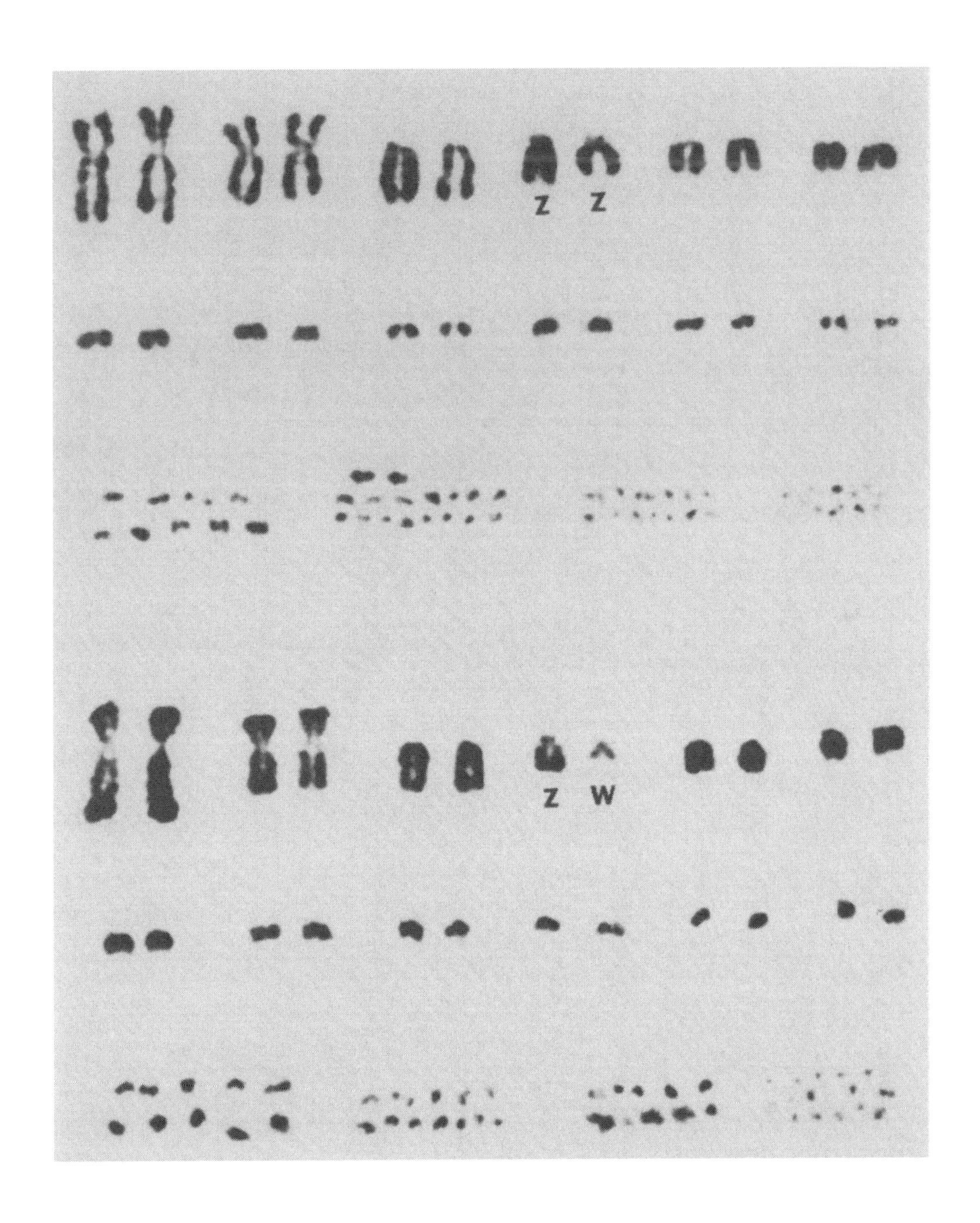

Order: ANSERIFORMES

Family: ANATIDAE
Tribe: ANATINI

Anas clypeata

(Common shoveler)

$2n = 78$

Order: ANSERIFORMES

Family: ANATIDAE
Tribe: ANATINI

Anas clypeata (Common shoveler)

2n = 78

AUTOSOMES: 4-6 Submetacentrics
The remainder are presumed to be acrocentric

SEX CHROMOSOMES: Z Appears to be acrocentric
W Acrocentric

Single pair specimens were supplied by Mr. Forrest Lee of the Northern Prairie Wildlife Research Station, U.S.F.W.S., Jamestown, N.D. Mitotic plates were obtained from feather pulp.

Anas clypeata (Common shoveler)

2n = 78

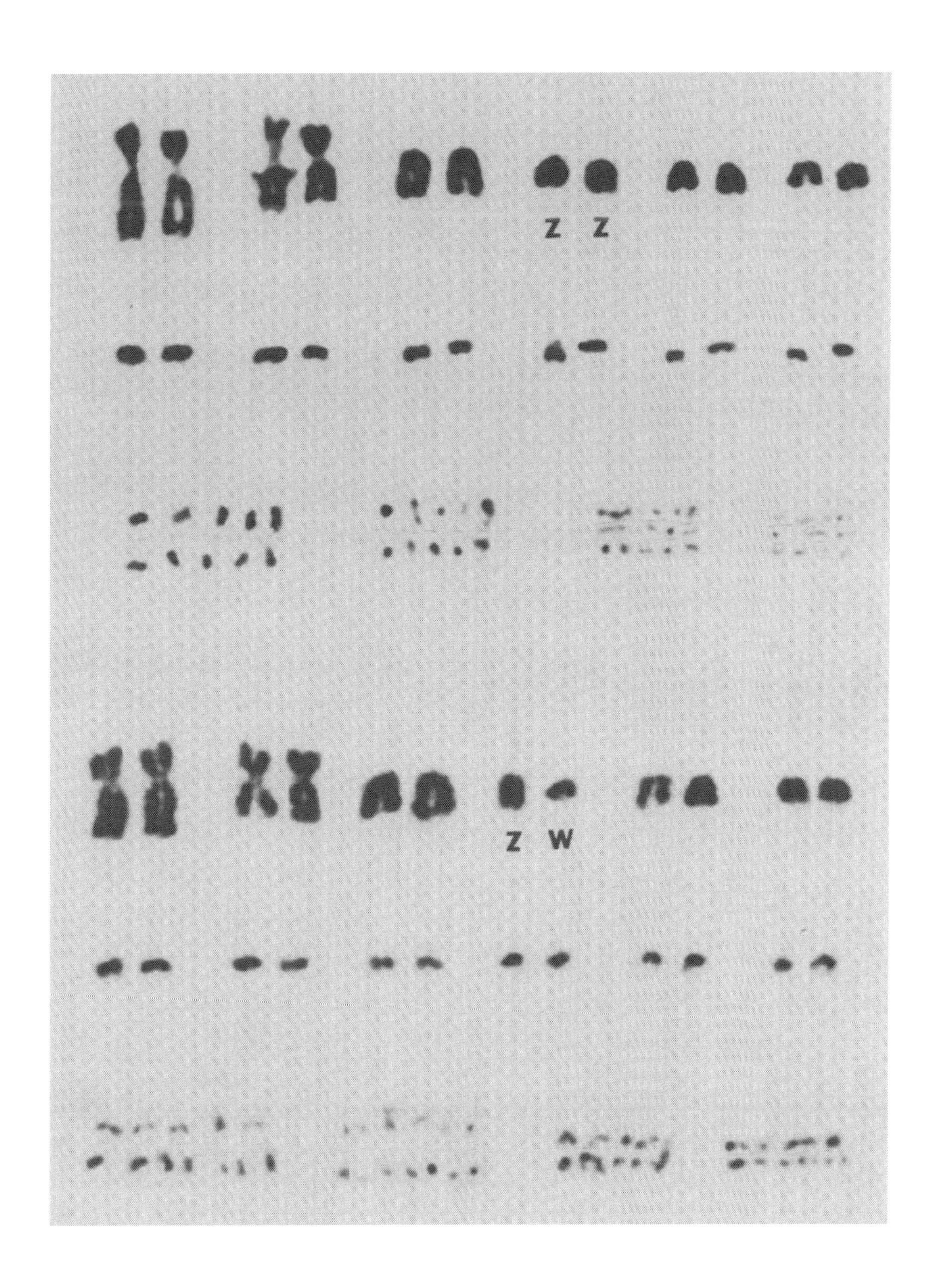

Order: ANSERIFORMES

Family: ANATIDAE
Tribe: ANATINI

Anas discors discors

(Prairie blue-winged teal)

2n = 80

Order: ANSERIFORMES

Family: ANATIDAE
Tribe: ANATINI

Anas discors discors (Prairie blue-winged teal)

2n = 80

AUTOSOMES: 4-6 Submetacentrics
The remainder are presumed to be acrocentrics

SEX CHROMOSOMES: Z Subtelocentric
W Acrocentric

Specimens of this species were supplied by Mr. Forrest Lee of the Northern Prairie Wildlife Research Station, N.S.F.W.S., Jamestown, N.D. Feather pulp was used as a source of mitotic plates.

Order: ANSERIFORMES

Family: ANATIDAE
Tribe: ANATINI

Anas discors discors (Prairie blue-winged teal)

2n = 80

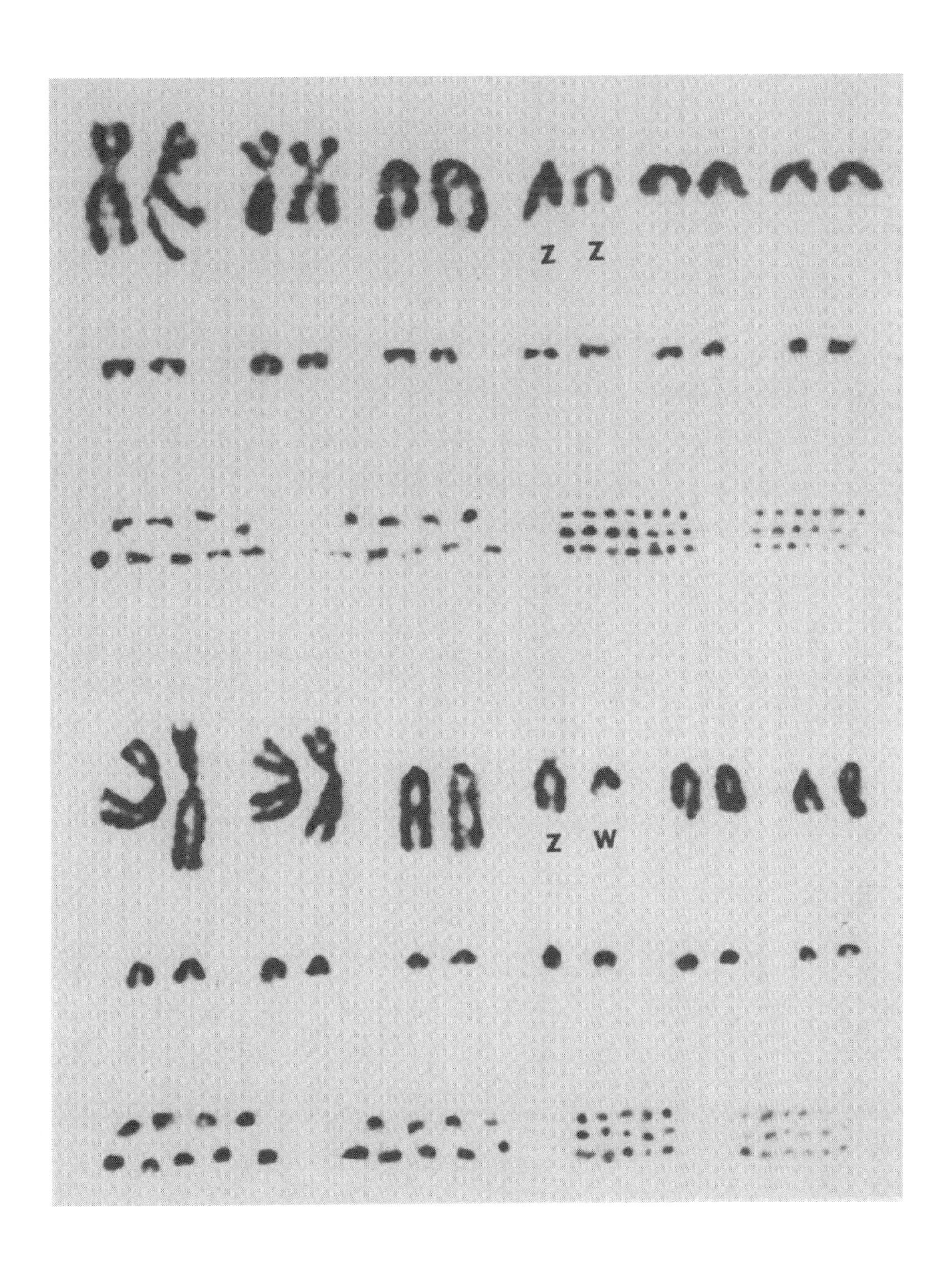

Order: ANSERIFORMES

Family: ANATIDAE
Tribe: ANATINI

Anas platyrhynchos platyrhynchos

(Mallard and its domesticated varieties)

$2n = 80$

Order: ANSERIFORMES

Family: ANATIDAE
Tribe: ANATINI

Anas platyrhynchos platyrhynchos

(Mallard and its domesticated varieties)

2n = 80

AUTOSOMES: 6 Submetacentrics
The remainder are presumed to be acrocentrics

SEX CHROMOSOMES: Z Subtelocentric
W Acrocentric

Several specimens of Mallards of the wild type, game farm varieties, and domesticated varieties, e.g., Pekin Duck, were obtained in the vicinity of St. Paul, Minnesota. Mitotic plates were obtained from feather pulp, embryonic tissue, and leucocyte cultures. The karyotype of the Mallard is typical of the Dabbling Ducks (*Anatini*) so far examined in this laboratory. The Z chromosome is 4th largest in size and is subtelocentric. It can be readily distinguished from numbers 5 and 6 in the better preparations. Hammar (1966) concluded that the Z chromosome is number 6.

REFERENCES:

1) Ohno, S., C. Stenius, L. C. Christian, W. Beçak, and M. L. Beçak. 1964. Chromosomal uniformity in the avian subclass *carinatae*. Chromosoma (Berl.) 15, 280-288.

2) Hammar, B. 1966. The karyotypes of nine birds. Hereditas 55, 367-385.

3) Mott, C. L., L. H. Lockhart, and R. H. Rigdon. 1968. Chromosomes of the sterile hybrid duck. Cytogenetics 7, 403-412.

4) Bloom, S. E. 1969. Mitotic chromosomes of Mallard ducks. J. Heredity 60, 35-38.

Order: ANSERIFORMES

Family: ANATIDAE
Tribe: ANATINI

Anas platyrhynchos platyrhynchos

(Mallard and its domesticated varieties)

2n = 80

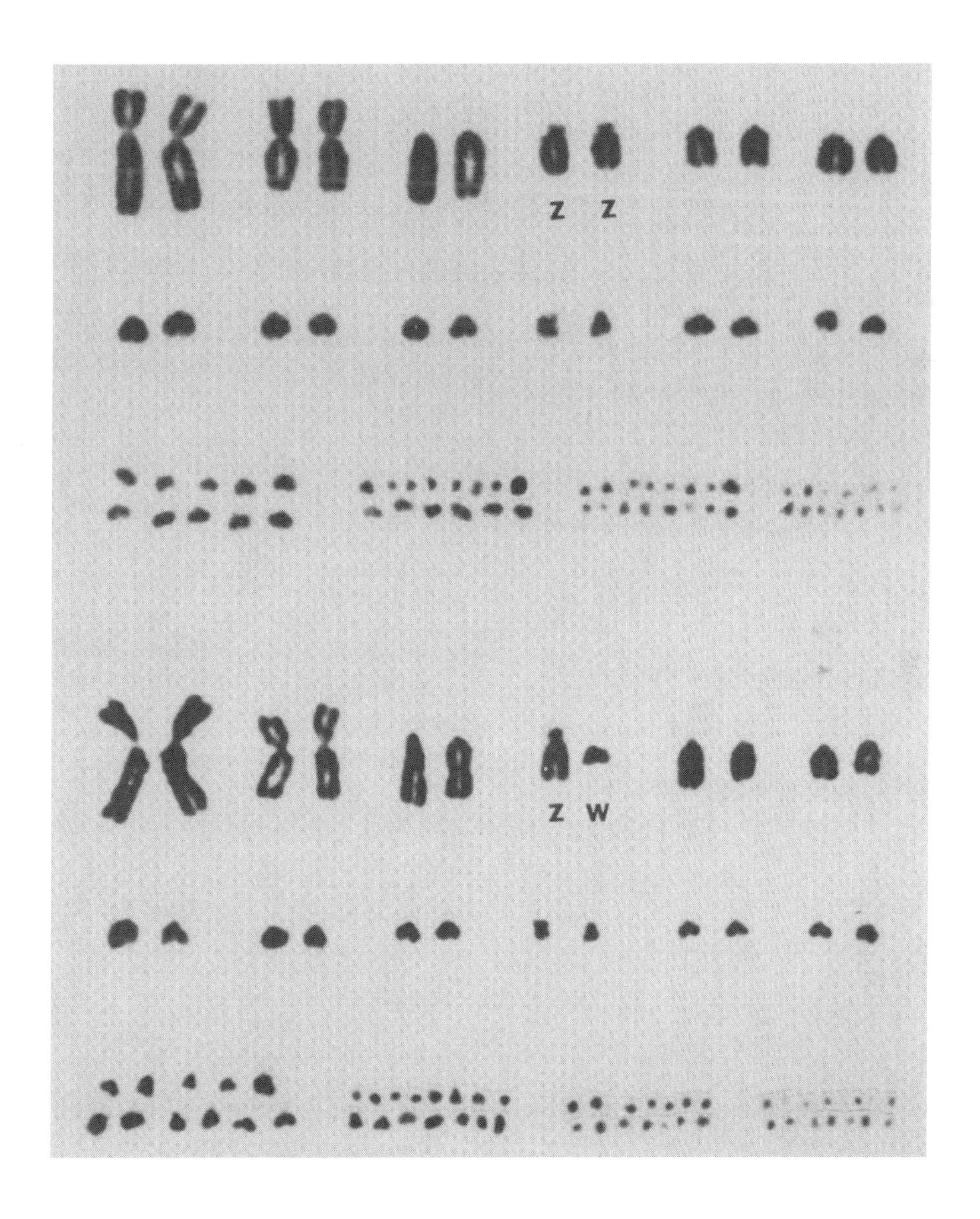

Order: ANSERIFORMES

Family: ANATIDAE
Tribe: ANATINI

Anas streptera

(Gadwall)

2n = 80

Order: ANSERIFORMES

Family: ANATIDAE
Tribe: ANATINI

Anas streptera (Gadwall)

2n = 80

AUTOSOMES: 4-8 Submetacentrics
The remainder are presumed to be acrocentrics

SEX CHROMOSOMES: Z Subtelocentric
W Subtelocentric or acrocentric

Specimens were supplied by Mr. Forrest Lee of the Northern Prairie Wildlife Research Station, U.S.F.W.S., Jamestown, N.D. Mitotic plates were obtained from feather pulp.

Order: ANSERIFORMES

Family: ANATIDAE
Tribe: ANATINI

Anas streptera (Gadwall)

2n = 80

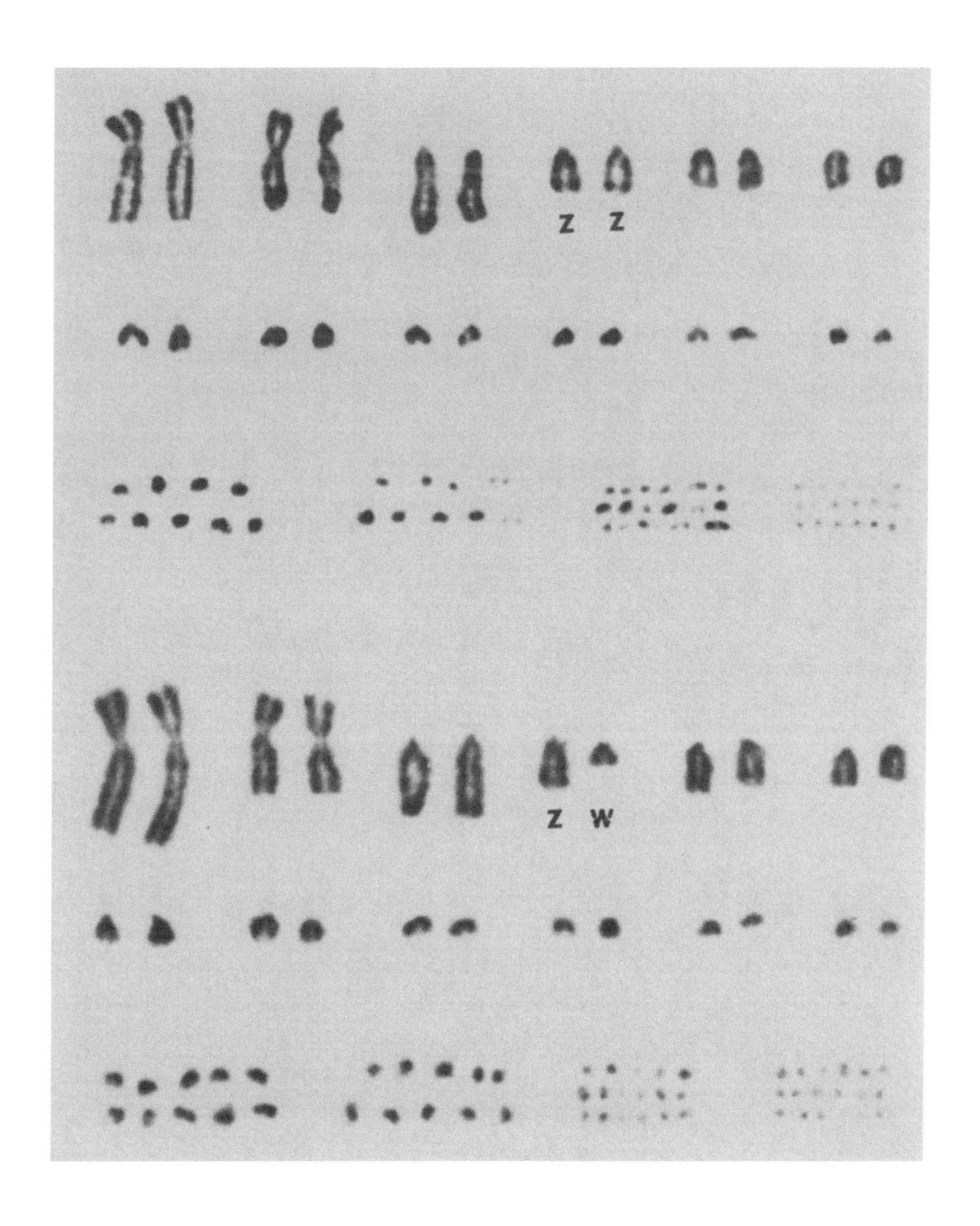

Order: ANSERIFORMES

Family: ANATIDAE
Tribe: AYTHYINI

Aythya valisineria

(Canvasback)

$2n = 80$

Order: ANSERIFORMES

Family: ANATIDAE
Tribe: AYTHYINI

Aythya valisineria (Canvasback)

2n = 80

AUTOSOMES: 6 Submetacentrics
The remainder are presumed to be acrocentrics

SEX CHROMOSOMES: Z Acrocentric
W Acrocentric or subtelocentric

Single pair specimens were supplied by Mr. Forrest Lee of the Northern Prairie Wildlife Research Station, U.S.F.W.S., Jamestown, N.D. Mitotic plates were obtained from feather pulp. The karyotype is indistinguishable from the Redhead (*Aythya americana*) (Folio Av-32) and Lesser Scaup (*A. affinis*) (Folio Av-31).

Order: ANSERIFORMES

Family: ANATIDAE
Tribe: AYTHYINI

Aythya valisineria (Canvasback)

2n = 80

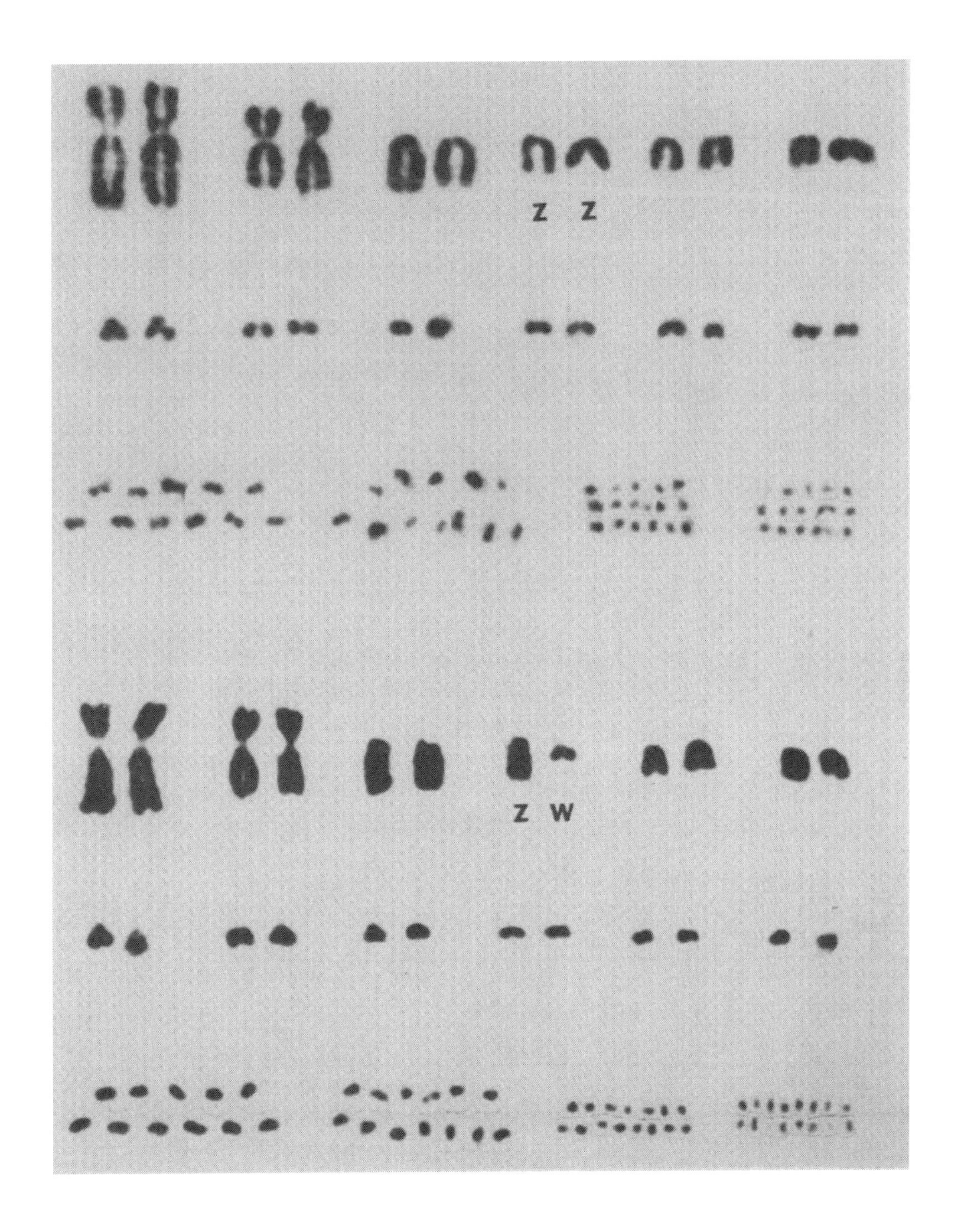

Order: ANSERIFORMES

Family: ANATIDAE
Tribe: AYTHYINI

Aythya affinis

(Lesser scaup)

$2n = 80$

Order: ANSERIFORMES

Family: ANATIDAE
Tribe: AYTHYINI

Aythya affinis (Lesser scaup)

2n = 80

AUTOSOMES: 6 Submetacentrics
The remainder are presumed to be acrocentrics

SEX CHROMOSOMES: Z Acrocentric
W Acrocentric or Subtelocentric

Single pair specimens were supplied by Mr. Forrest Lee of Northern Prairie Wildlife Research Station, U.S.F.W.S., Jamestown, N.D. Mitotic plates were obtained from feather pulp tissue. The karyotype is similar if not identical to that of the Canvasback (*A. valisineria*) (Folio Av-30) and Redhead (*A. americana*) (Folio Av-32).

Order: ANSERIFORMES

Family: ANATIDAE
Tribe: AYTHYINI

Aythya _affinis_ (Lesser scaup)

2n = 80

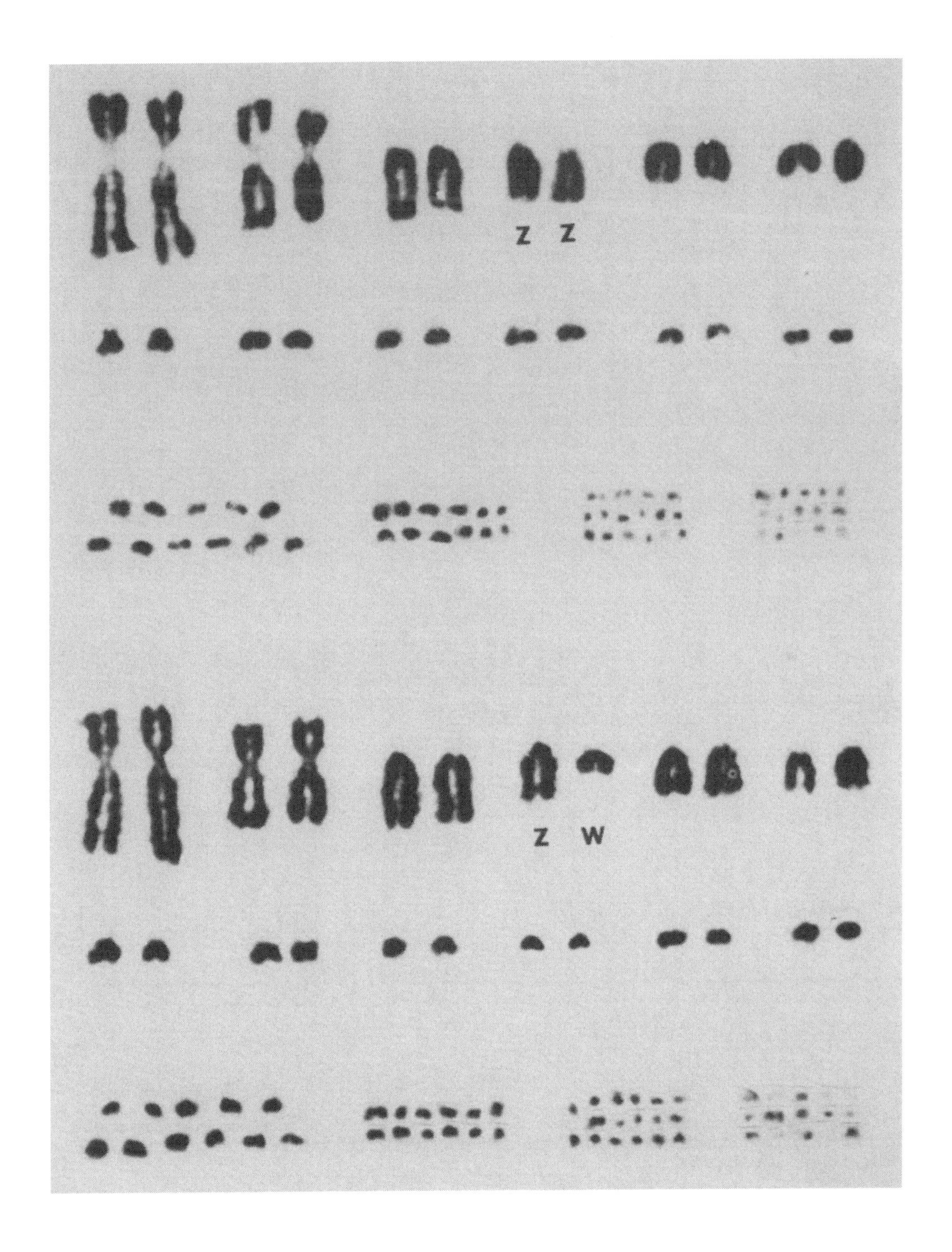

Order: ANSERIFORMES

Family: ANATIDAE
Tribe: AYTHYINI

Aythya americana

(Redhead)

$2n = 80$

Order: ANSERIFORMES

Family: ANATIDAE
Tribe: AYTHYINI

Aythya americana (Redhead)

2n = 80

AUTOSOMES: 6 Submetacentrics
The remainder are presumed to be acrocentrics

SEX CHROMOSOMES: Z Acrocentric
W Acrocentric or subtelocentric

Single pair specimens were supplied by Mr. Forrest Lee of the Northern Prairie Wildlife Research Station, U.S.F.W.S., Jamestown, N.D. Mitotic material was obtained from feather pulp tissue. The karyotype is similar if not identical to that of the Canvasback (*A. valisineria*) (Folio Av-30) and the Lesser Scaup (*A. affinis*) (Folio Av-31).

Order: ANSERIFORMES

Family: ANATIDAE
Tribe: AYTHYINI

<u>Aythya</u> <u>americana</u> (Redhead)

2n = 80

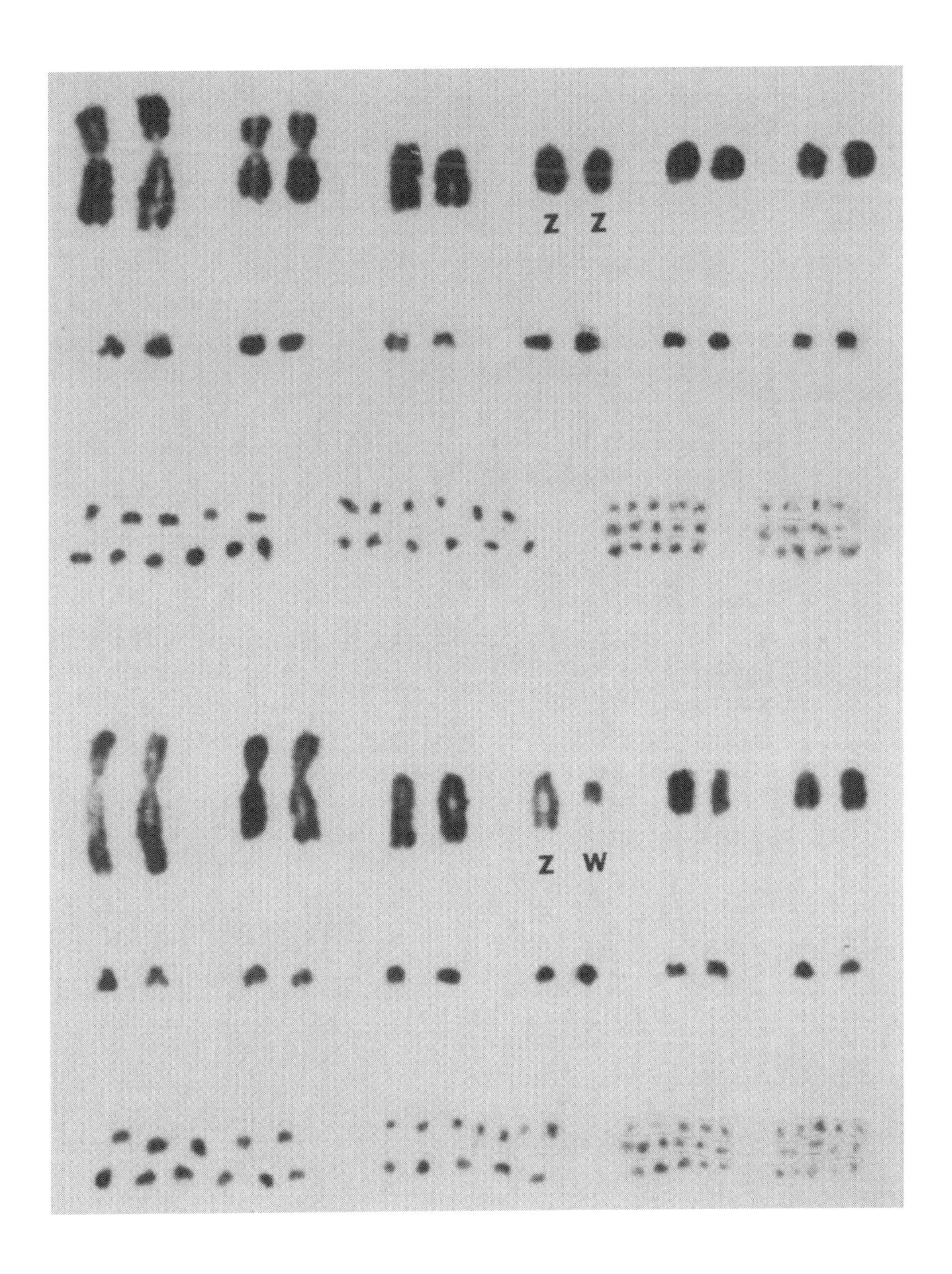

Order: ANSERIFORMES

Family: ANATIDAE
Tribe: TADORNINI

Tadorna tadorna

(Common shelduck)

2n = 80

Order: ANSERIFORMES

Family: ANATIDAE
Tribe: TADORNINI

Tadorna tadorna (Common shelduck)

2n = 80

AUTOSOMES: 10 or more metacentrics, submetacentrics or subtelocentrics
The remainder are presumed to be acrocentrics

SEX CHROMOSOMES: Z Acrocentric
W Acrocentric

Single pair specimens were supplied by Mr. Forrest Lee of the Northern Prairie Wildlife Research Station, U.S.F.W.S., Jamestown, N.D. Mitotic plates were obtained from feather pulp.

Hammar (1970) suggests that the chromosome number 5 is probably the Z chromosome in this species. Because of the similarity in size of chromosome numbers 4, 5, and 6, it is difficult to determine which one is definitely the Z chromosome. We believe that chromosome number 4 is the Z chromosome, based on measurements made on the chromosomes of the heterogametic female.

REFERENCES:

1) Hammar, B. 1970. The karyotypes of thirty-one birds. Hereditas 65, 29-58.

Order: ANSERIFORMES

Family: ANATIDAE
Tribe: TADORNINI

Tadorna tadorna (Common shelduck)

2n = 80

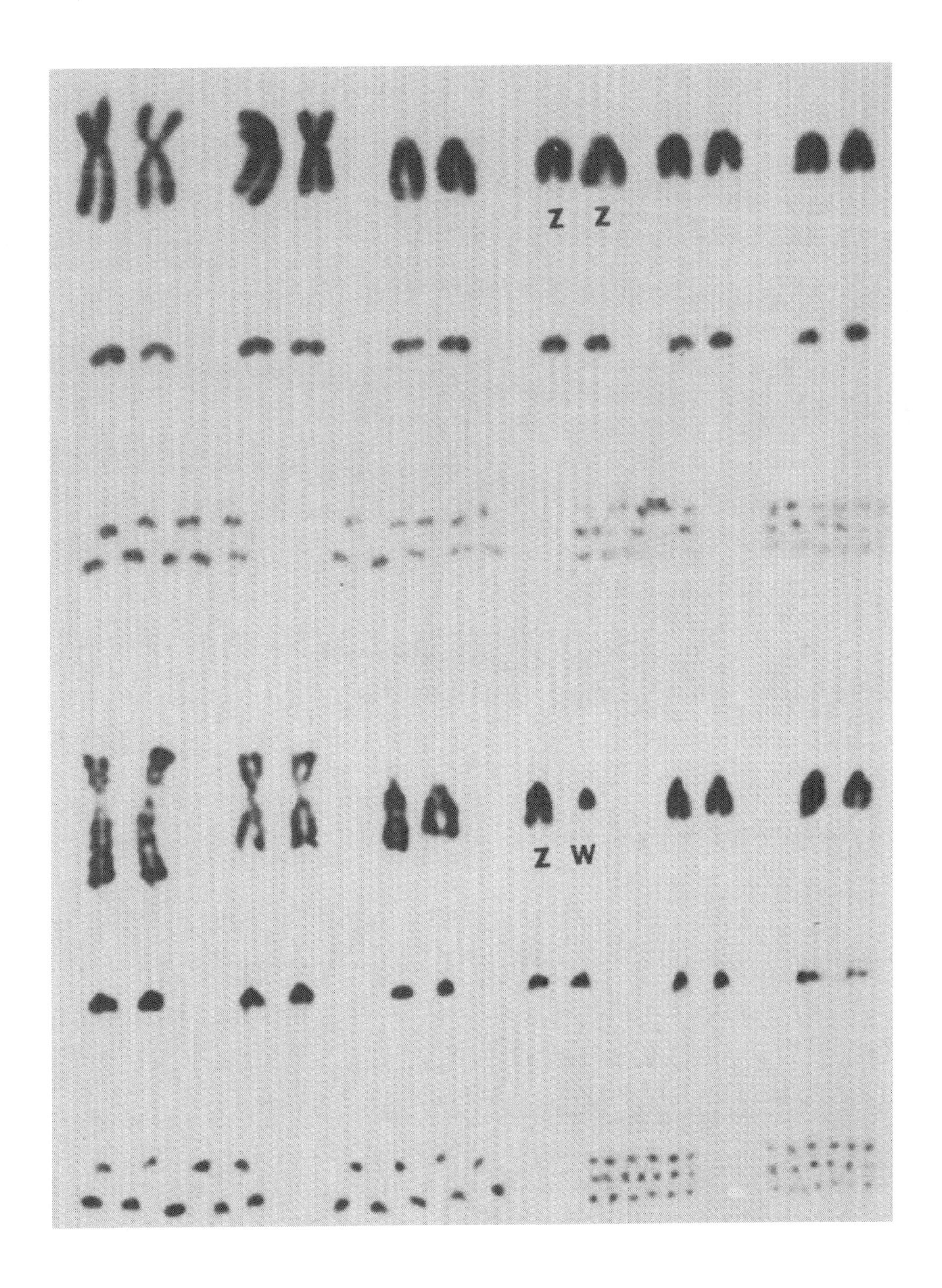

Order: ANSERIFORMES

Family: ANATIDAE
Tribe: MERGINI

Bucephala clangula americana

(American goldeneye)

2n = 80

Order: ANSERIFORMES

Family: ANATIDAE
Tribe: MERGINI

Bucephala clangula americana (American goldeneye)

2n = 80

AUTOSOMES: 14 Subtelocentrics
The remainder appear to be acrocentric

SEX CHROMOSOMES: Z Subtelocentric
W Subtelocentric

Single pair specimens were kindly furnished by Mr. Forrest Lee of the Northern Prairie Wildlife Research Station, U.S.F.W.S., Jamestown, N.D. Feather pulp was used as a source of mitotic plates.

This species presents an unusual karyotype among birds, as all macrochromosomes including the Z and W are subtelocentric. The Z chromosome is very similar in size to numbers 5 and 6 and difficult to distinguish. The size relationship of the W to the Z chromosome is large as compared to other avian species. The W is hard to distinguish from chromosomes 7 and 8. Hammar (1970) states that the number 3 chromosome is the Z.

REFERENCES:

1) Hammar, B. 1970. The karyotypes of thirty-one birds. Hereditas 65, 29-58.

Order: ANSERIFORMES

Family: ANATIDAE
Tribe: MERGINI

Bucephala clangula americana (American goldeneye)

2n = 80

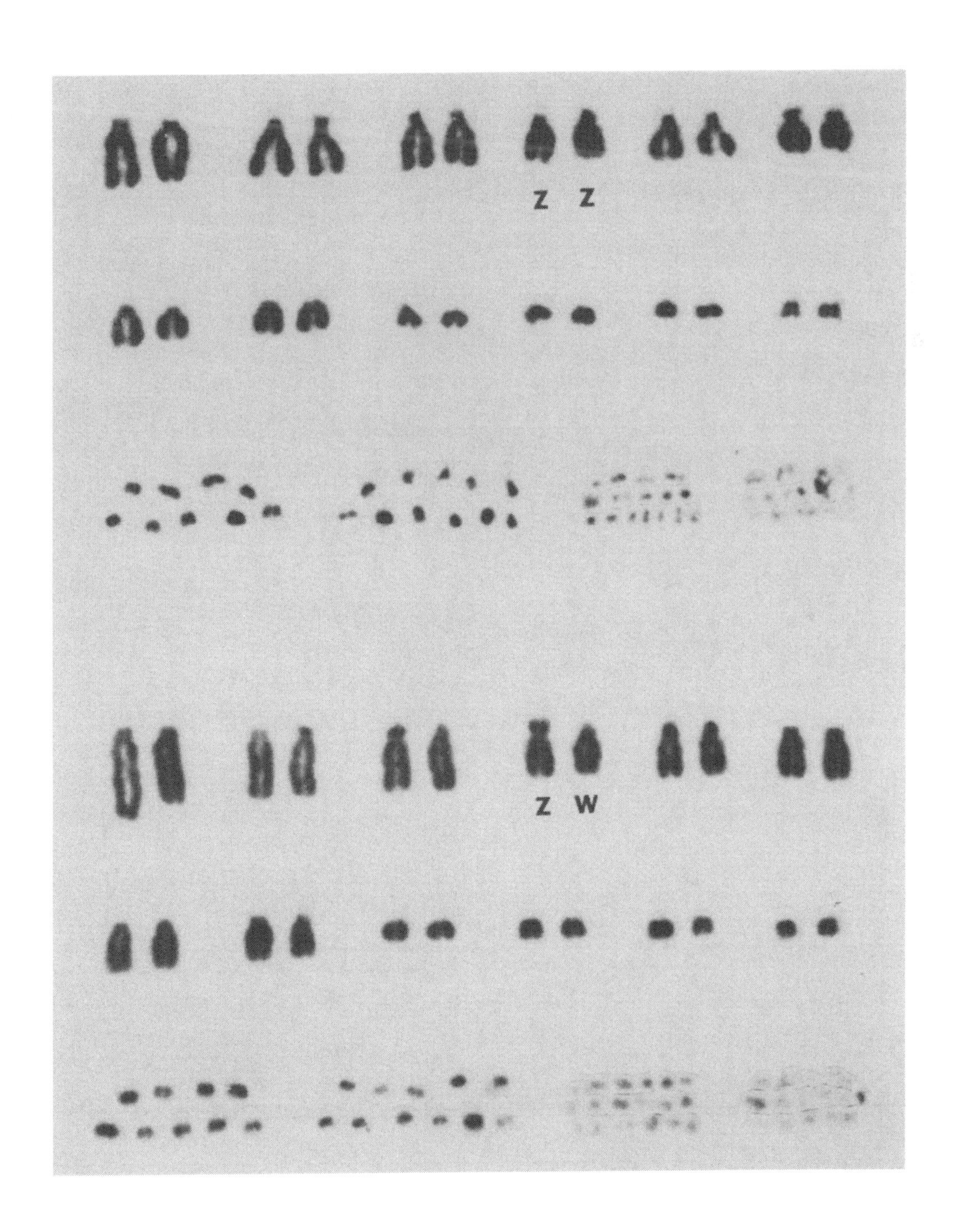

Order: SALMONIFORMES

Family: SALMONIDAE

Salvelinus namaycush

(Lake trout)

2n = 84

Order: SALMONIFORMES   Family: SALMONIDAE

Salvelinus namaycush (Lake trout)

2n = 84

CORRIGENDUM

Re: Folio P-6, 1971, *Salvelinus namaycush* (Lake trout)

This species belongs to the Order Salmoniformes, not the Clupeiformes as listed. The reason for this listing is the change in taxonomic listings between the second and third editions of the American Fisheries Society's: "A List of Common and Scientific Names of Fishes from the United States and Canada," whose nomenclature is adopted.

Salvelinus namaycush (Lake trout)

2n = 84

Index, Volume 2

<u>Tropidurus torquatus</u>	R-29
<u>Xenodon neuwiedii</u>	R-23

Cumulative Index, Volumes 1 & 2

Cumulative Index, Volumes 1 & 2